ALAMANCE COMMUNITY COLLEGE--LRC
00026015

QK
31
.W345
T76
1993

Troyer, James R.
Nature's champion

ALAMANCE
COMMUNITY COLLEGE

GAYLORD

Nature's Champion

JAMES R. TROYER

Nature's Champion

B. W. Wells, Tar Heel Ecologist

The University of North Carolina Press
Chapel Hill & London

© 1993 The University of North Carolina Press
All rights reserved
Manufactured in the United States of America

The paper in this book meets the guidelines for permanence and durability of the Committee on Production Guidelines for Book Longevity of the Council on Library Resources.

Publication of this book was aided by a subvention from the Department of Botany at North Carolina State University.

frontispiece: B. W. Wells in about 1940. Courtesy of Maude Barnes Wells.

Library of Congress Cataloging-in-Publication Data
Troyer, James R.
Nature's champion : B. W. Wells, Tar Heel ecologist / by James R. Troyer.
p. cm.
Includes bibliographical references and index.
ISBN 0-8078-2081-4 (alk. paper)
1. Wells, B. W. 2. Botanists—North Carolina—Biography. 3. Ecologists—North Carolina—Biography. I. Title.
QK31.W345T76 1993
581'.092—dc20
[B] 92-43554
CIP

97 96 95 94 93 5 4 3 2 1

TO LARRY ALSTON WHITFORD

CONTENTS

ILLUSTRATIONS

ACKNOWLEDGMENTS

I FIRST FACED B. W. Wells one spring day in 1957 when I attended a seminar at North Carolina State College. As a young assistant professor I went to that seminar early, was the first to arrive, and chose a comfortably out-of-the-way seat, prepared to doze through enlightenment. Suddenly an old fellow whom I had never seen before burst in, scanned the almost empty room, marched directly over to me, and sat down with authority. At once he fixed me with sparkling eyes and launched into a long and vigorous speech: the Carolina Bays, he declaimed, were obviously the result of a meteorite shower, there was convincing evidence for this inescapable conclusion, and only a fool would think otherwise. I was nonplussed. I thought that maybe he had mistaken me for someone else or perhaps was simple-minded. New to the region, I had no idea what a Carolina Bay was; for all I knew it might have been a kind of horse. Although I did not know what he was talking about, at some point in his lecture—for that is what it was—I realized that he was making me *want* to know what he was talking about. Later I understood that on that day I had been in the presence of a Teacher and learned that his name was B. W. Wells.

He was retired when I knew him, but he came around to his old college department fairly often over the years, usually to bring in a fresh painting to display in the department office. On such visits he would stop for a brief conversation. Although our lives and interests touched only occasionally and peripherally, those little conversations were never uninteresting, usually stimulating, sometimes deeply challenging, and always unforgettable. So my thanks first of all must go to B. W. Wells himself, or at least to his memory, simply for having been.

The bone of this account lies in the public record, to which references are duly made. The flesh, however, grew from the recollections of many kind persons. Final realization of it was made possible by assistance from numerous other quarters. For all of this help sincere thanks are expressed.

The Department of Botany of North Carolina State University

and its faculty, through the Botany Enhancement Fund, munificently provided a subvention for the publication of this work.

Other organizations and institutions provided information or access to information. Thankful acknowledgment is made to the Interlibrary Loan Department of the D. H. Hill Library of North Carolina State University, the Archives of North Carolina State University, the University of North Carolina Press, the Manuscript Department of the University of North Carolina Library, the Garden Club of North Carolina, Inc., the B. W. Wells Association, Inc., and the North Carolina Wild Flower Preservation Society, Inc.

Numerous individuals freely gave information or assistance of various kinds, some in large, some in small, many in more than one, but all in important ways. Somewhat perversely, perhaps because my own name has always appeared near the ends of lists, grateful acknowledgment is made to these helpful persons in reverse alphabetical order: Donna S. Wright, Larry A. Whitford, Duella Whitford, Maude Barnes Wells, Ann B. Ward, Patricia Broomhall Troyer, Maurice S. Toler, Ann S. Smith, Ernest D. Seneca, H. T. Scofield, Ruth Reichard, Thomas L. Quay, G. Ray Noggle, Michael G. Martin, Jr., Ron Maner, John Lawrence, Mrs. W. C. Landolina, W. Benson Kirkman, H. Thomas Kearney, Matthew Hodgson, James W. Hardin, Charlotte Hilton Green, Joy Downs, Lisa M. Dellwo, Arthur W. Cooper, John M. Clarkson, Katherine Metcalf Browne, Stephen G. Boyce, Hardy D. Berry, Donald B. Anderson, Charles E. Anderson, and David A. Adams.

Thanks are also expressed to individuals who generously contributed to the Botany Enhancement Fund at the time of an appeal for this special purpose. They include, also in reverse alphabetical order, Richard S. Winder, Larry A. Whitford, Janet Hatley White, William F. Thompson, Mrs. George T. Stronach, Ernest D. Seneca, Mary Beth Pelletieri, Jimmie R. Overton, G. Ray Noggle, Linda M. Lamm, Willy Kalt, Mrs. Sydnor M. Cozart, Arthur W. Cooper, Stephen G. Boyce, Helen A. Boyce, Richard R. Bounds, and five anonymous donors.

Persons and organizations providing information did their best to make it accurate. Any errors of fact or interpretation propagated by this account must therefore be attributed to me.

J. R. T.
July 1992

PROLOGUE

WHO WAS Bertram Whittier Wells?

Answering this question superficially is easy, for the bare facts of his life can be quickly recited. He was born in Ohio in 1884 and grew up there. Having decided to become a botanist, he earned degrees from the Ohio State University and the University of Chicago. His career was academic, beginning with brief teaching stints at five colleges. In 1919 he joined North Carolina State College, now North Carolina State University, serving as head of the Department of Botany and Plant Pathology until 1949 and as professor until 1954. His early scientific studies on insect galls brought him national and international attention. In North Carolina, however, he drastically changed his research interests and became a significant pioneer in the young science of ecology; this work was his principal professional legacy. He actively popularized the plants and vegetation of the state and championed their preservation for all people to enjoy. This advocacy was crowned by his book, *The Natural Gardens of North Carolina*, published in 1932 and reprinted in 1967. As an educator he was outstanding in the teaching of students, in working for the high academic standards of a true university, and in standing publicly in defense of intellectual freedom. He married twice but had no children. He had more than the usual share of friends and admirers but also some detractors and enemies. After his professional retirement he moved to a primitive rural home where for twenty-four years he pursued an interest in art as an active painter and remained a voice for appreciation and conservation of nature. He died in 1978.

Facts about Wells are seldom bare, however; thus the question of who he was is not so easily disposed of in a deeper sense. Like all humans, he was a unique individual. Part of his uniqueness lay in an unusual combination of personal qualities and abilities. But part of it was the way these individual characteristics were cast into particular coordinates of time and space: he was very much a product of his age and venue. He was a gifted specimen of the upwardly mobile, supremely confident, thoroughly optimistic, middle-class, middle-American of the first half of the twentieth century. With an ideal of progress and a dream of success through talent and hard work, the usual such American could not see the less fortunate consequences of unbridled development, could scarcely imagine the constraints of limited possibility and restricted resource which we must face today. Wells could. Transfigured by the dazzling array of vegetation he met in his adopted state, he not only embraced the science of ecology but absorbed into his deepest being its implications for the everyday life of everyone. So he not only witnessed nature with clear eyes but also bore that witness loudly at every opportunity. Like the paladins of old who swore undying fealty to their sovereign, he championed the cause of nature with steadfast loyalty.

Wells was multifaceted. Although first of all a significant scientist, he also gave importantly and unselfishly to North Carolina, to his academic institution, and to an unusually large number of persons on whom his life impinged. Examining that life is like dissecting an onion, for he existed in a surprising number of different but interconnected spheres; peeling one away reveals others not previously evident but no less interesting. At the center, determining his actions in all the spheres, was the essential nature of the man himself. Because he was many-sided, the course of his life was not a simple single line, but rather contrapuntal spirals touching repeatedly on a number of motifs. Therefore his story is told here not chronologically but thematically: the question of who Bertram Whittier Wells was is to be answered in many ways.

PART I

Scientist

CHAPTER 1

Ecological Awakening

In the spring of 1920, on a visit to Wilmington, out of the railroad car window, I saw a vast flat area literally covered with wild flowers. I immediately made up my mind to see it again. . . . I became convinced there was no such area of equal size and perfection with over a hundred species of herbaceous wild flowers blooming in profusion from late February to middle December. . . . As my memory goes back over the 46 years in North Carolina the two summers of day after day on the Big Savannah continually surrounded in floral beauty while engaged in our technical soil studies stand out beyond everything else.—Bertram Whittier Wells, 1967

HE COULD HARDLY believe it. Spreading for half a mile before his eyes were wildflowers, a carpet of wildflowers over the treeless expanse of delicate green spring grass which mostly hid the black soil beneath. Even from the window of the railway car he recognized the myriad patches of white violets, the bright blue polka dots of irislike blooms, the nodding heads of what looked like white dandelions. But there were also flowers he did not know: huge golden goblets flung in groups on countless cloths of white fleabane, other tables set with glasses of red wine, still other whites, yellows, and blues, all strange and magnificent. The view across the car to the other side was the same and just as vast. As the leisurely train traversed this banquet of floral delights, he estimated its length at two miles. Incredible! At least fifteen hundred acres of wildflowers were thrusting up their multicolored beauty in a dazzling array. It was like a huge garden thickly planted on a well-kept lawn. But it was not planted, it was natural, a natural garden. He was thirty-six years old and a professional botanist, but he had never met a floral display of such size and perfection: not in the Midwest, not in Kansas, Arkansas, or Texas, certainly not in Ohio, his birthplace. Immediately he knew that he must see it again, walk through it, breathe more slowly and more deeply its beauty and wonder. Not just see it, but study it, ponder it, explain it. He sensed the fascinating problems it contained: why was it treeless? why were there so many flowers? why was it just here in this obscure cranny of eastern North Carolina? He knew he had to return, he had to find out.[1]

The floral wonder which the Atlantic Coast Line Railroad bisected was the Big Savannah, an unusually extensive upland grass-sedge bog, a plant community dominated by grasses, sedges, and numerous showy herbs. Occurring as localized vegetational dots on the southeastern coastal plain, such a community may or may not include pine trees. When the pines are present, they are usually scattered, giving the locality the aspect of a savannah; when, as is often the case, fire or human activity has removed the pines, the aspect is that of a small prairie. Widely known among the inhabitants as "savannahs," these features have also been called "pine flat woods" or "pine barrens" if the trees are present and "open grounds" if they are not. Located about two miles north of Burgaw in Pender County, the Big Savannah was a remarkable specimen; the treeless portion of it was more than two miles long and about one mile wide.[2]

To the questioning scientific mind of Bertram Whittier Wells, this chance first encounter with the Big Savannah was crucial: it changed his life completely. He was a professor who had just come the previous autumn to the South and to North Carolina State College in Raleigh. As a botanist in 1920, he was much more conversant with the broad aspects of his science than is any botanist today, for the simple reason that there was less to know; the immense eruption of scientific knowledge with which present-day scientists contend had not yet occurred. But scientific research requires one to be narrowly specific, sometimes agonizingly so, if fruitful results are to be achieved. Therefore, although his outlook was broad and his knowledge deep for the time, he had concentrated his own technical studies on plant galls, those bits of localized abnormal growth on plants caused by insects or other such agents. The Big Savannah changed that. It opened his eyes in an intense way to the problems of plant ecology, the science of plants in relation to their surroundings. Until then he had known about ecology in the rather distant way of an academic scientist whose personal excitement is generated by other things. It had not been his specialty at the University of Chicago, but he had studied with a world leader in that field, Henry Chandler Cowles. Both in Chicago and at the Ohio State University he had also absorbed the ideas of other ecologists, especially Edgar Nelson Transeau, Paul Bigelow Sears, and George Damon Fuller. Now, aroused and excited by the Big Savannah, he looked about his new home state of North Carolina and found an array of vegetation and ecological problems which awed his mind, captured his heart, and directed his professional work from that time on. He would study his galls for a few years to come, but he would do so with the calm fondness one feels for a youthful love the fire of which has been banked. His heart was now elsewhere: the Big Savannah had stirred into life the ecologist that was in him.

The plant ecology in which Wells enlisted in 1920 was a very young science. In the second half of the nineteenth century botanists had begun to expand their horizons far beyond previously traditional studies of the form and structure of plants (morphology and anatomy) or of their classification and naming (taxonomy and systematics). New approaches emphasized dynamic aspects such as their processes and functions (physiology), inheritance of their characteristics (genetics), and relationships to their surroundings or environments (ecology). Plant ecology grew out of the older science of the distribution of plants and vegetation on the earth (plant geography). It went beyond merely describing such distributions to explaining their causes in terms of environmental factors such as climate, soil type, configuration of the land, water relations, and the presence of other living things. It was a desire to participate in this explanation of why plants grow where they do which erupted in Wells in response to the challenge of the Big Savannah. Although a single plant such as an oak tree can be considered in relation to its surroundings, the emerging science of plant ecology soon began to stress the relations of plants in natural groups, or communities. A community, for example a forest dominated by oak and hickory trees, is an entire set of plants that live together in the same place. Plant ecologists early on believed they could discern a correspondence between sites or habitats and communities; over a fairly broad region of the landscape a given type of plant community tended to occupy the same kind of place. A fundamental question, one of those posed to Wells by the Big Savannah, was why this correspondence existed.[3]

Although it had its roots in nineteenth-century European science, plant ecology in the United States began to be a recognizable branch of botany only after about 1900. As it developed in this country, the new subdiscipline studied both the makeups and the locations of plant communities, as its forerunners had. It also considered the classification of communities and the terminology needed to describe them. But it put especially strong emphasis on sequential changes over time in the communities occupying a given place. This dynamic process, known as plant succession, typically involves an orderly series of different communities occurring one after the other until a stable, self-reproducing one appears which is in long-term balance with the surroundings. Such successional change is propelled by alterations in the habitat produced by each community in its turn, alterations that then permit a new community to become established and develop. The climax community, the stable one at the end of the successional line, was thought by many early ecologists to be in delicate balance with the climate of the region and thus determined by it. This emphasis on successional dynamics was a major thrust of the American plant ecology which Wells joined.[4]

Wells conducted professional ecological research from 1920, when he encountered the Big Savannah, until 1954, when he retired from his professorship. He thus worked during the first half of the twentieth century, a period recognized by ecologists as a fairly distinct initial stage in the development of their science. American professionals producing significant amounts of ecological work entirely during this period numbered roughly about thirty men and several women. This first generation of plant ecologists was characterized by distinctive methods and outlook. They simply went out into the field and noted carefully what they saw there; they generally recorded their observations in words, not numbers; and they inferred the dynamics of change with time by drawing conclusions from what they perceived in different places. Like the pioneers in most fields of science, they produced results that were descriptive and verbal, not quantitative and mathematical. As might be expected, sometimes this nonquantitative approach led them into error. But by and large they were a shrewdly observant and carefully logical lot, and most of the results they described correctly formed the foundation upon which later workers built.[5]

Wells was truly one of these first-generation ecologists. The fascination of work with plants in the natural outdoors had lured him into botany at an early age, and in 1920 he was a mature and accomplished field observer. His subsequent ecological studies revealed that he could discern relations and processes on the landscape that would have eluded most persons. In the words of a later ecologist, he confronted the world directly with "a pencil, a paper, and a brain." Like others of his generation, he usually recorded the kinds or species of plants making up a community, laying emphasis on the dominant ones. To an ecologist, dominant species are those that determine the character of a community because of their large sizes or predominant numbers; plant communities are usually named for their dominant species, as in the oak-hickory forest or the cattail marsh. But Wells's approach was *ecological*. He not only described communities, he also tried to relate their distribution through space and time to factors of their surroundings such as soil properties, atmospheric conditions, and topography of the land. In all of these matters he sought to uncover causes and effects. He described what he saw, but, more important, he offered explanations of what he saw.[6]

Wells's field techniques were those common to plant ecologists of the first generation. He often made observations by examining a community along a transect, a sampling line carefully chosen to reveal the characteristics of the site. On one occasion he studied plants occurring in quadrats; these are square or rectangular sampling areas marked out within a community so that representative observations can be made in exactly the same places at different times. He usually reported the composition of a community by simply listing the names of the plant species he found in it; often he indicated the relative numbers of different plants by subjectively assigning them to arbitrary frequency classes. Factors of the habitats in which communities grew were often measured with analytical techniques of physics and chemistry. The factors examined included such atmospheric conditions as humidity, temperature, brightness of light, and evaporation rate, as well as such soil conditions as temperature, texture, acidity, and water-holding power. In a few cases he studied ecological plant anatomy, a subject popular during the early period of ecology. Such an approach involved trying to correlate variations in the internal tissues of plants with the habitats in which they grew. In several studies he carried out what the early ecologists called transplant, potted-plant, and phytometer studies. These involved moving plants from one community to another, potting up crop and natural plants in soils brought in from the field, and setting out crop plants in the habitats of various natural communities. It is interesting that in Wells's research technical operations such as chemical measurements, studies of anatomy, and transplanting procedures were done only when he had the help of a co-worker, most notably when that co-worker was Ivan Vaughan Detweiler Shunk, a soil bacteriologist by background. Wells was a field observer par excellence, not a laboratory technician.[7]

Because he was primarily oriented toward outdoor work in field and forest, Wells did not usually carry out experiments, studies in which scientists test proposed explanations by varying and controlling conditions to find out what then happens. But on several occasions he did perform extremely simple experiments to provide support for ideas based on field observations. In one case he noticed in the field that drainage water from humus-rich savannah soil was always perfectly clear, whereas that from sandy soil was always dark in color. He then established a correlation by simply passing some dark sandy-soil water through a sample of savannah soil and finding that it came out clear. In another instance he observed that in the hot, dry conditions in the sandhills the leaves of young turkey oak seedlings were almost always oriented vertically. Using wire, he forcibly held some leaves in a horizontal position and found that they quickly developed high-temperature injury from the greater amount of sunlight that struck them. After he discovered through field observations that injury to plants growing near the ocean was caused by windblown salt spray and not wind alone, he was able to duplicate such injury exactly on plants away from the ocean by misting them with seawater or salt solutions, using small medicinal sprayers purchased from a Wilmington drugstore. When he studied the lakes in Bladen County, he came to suspect that their fine bottom silt was brought in by wind rather than by water. He obtained experimental support for this idea by pouring some coastal-plain sand in front of an electric fan and collecting the material that blew across the room; it turned out to be indistinguishable under a microscope from the silt of the lakes. And during the same investigation, he concluded in the field that water was moving away from White Lake by an outward underground flow. He demonstrated the correctness of this conclusion by digging a series of small wells in the ground at various distances from the shore; when a dye was added to the well nearest the lake margin, it later appeared sequentially in those farther and farther away. These experiments were all simple. But the test of good science is not complexity; it lies in whether what is done gets at the heart of a matter. Wells's little experiments were all aimed precisely at crucial points and are thus demonstrative of his high level of scientific competence.[8]

Although the plant ecologists of the first generation used similar approaches and techniques, they differed greatly in the mental frameworks by which they organized the knowledge gained from field research. Such theoretical differences led to an unfortunate emphasis on doctrine, resulting in what a later ecologist called "The Age of School-asticism." During this period the acceptance of scientific ideas depended not only on facts and hypotheses but also on belonging to a network of collaboration, on having work acknowledged rather than ignored, and even on placing students in strategic jobs. Although there were several schools of thought, the one that dominated American plant ecology until about 1950 was led by Frederic Edward Clements. Clements's conceptual system included a detailed classification of plant communities into a rigid hierarchy of vegetational units. Also, the Clementsian system regarded vegetation as an actual organism that grows and develops through the succession of communities until it matures as a climax community. Because this climax was supposed to be determined by climate, it was presumed to be the same over a very large geographical region. The monoclimax theory, this idea of a single climatically determined stable community, proved ultimately to be a cloud over American ecological thought, but during the early period it predominated. As ecologists gained more and more facts, however, they gradually uncovered many cases of plant communities that were stable in time, and therefore climaxes, but were preserved by forces other than climate: soil conditions, factors of topography, even recurrent fire. Clements was then forced to add more and more special or exceptional categories to his classification system to preserve the sanctity of the climatic monoclimax. As a result, his system became unwieldy and artificial.[9]

Wells came to know Clements personally in 1925, when he sought advice in designing his detailed study of the Big Savannah community. In his earlier publications he classified the communities he studied into the Clementsian categories. But after about 1937 he never again made use of them; when he referred to a vegetational unit, he called it simply a *community*. He also initially swallowed the Clementsian monoclimax notion, which held that the climax for the Southeast as well as for the entire eastern United States was the beech and maple forest. But his own observations after about 1930 convinced him that there was essentially no such prominent beech-maple community present in the coastal plain and that other communities were so stabilized by factors other than climate that they, too, should be considered climaxes. After he discovered and described in 1939 climaxes on the coast determined by the salt-spray factor, he quickly and openly assumed a polyclimax view; he had seen with his own eyes not one, but more than one, essentially permanent communities. Some of Wells's associates sometimes criticized him for being too Clementsian. Certainly he initially constrained his thought within that system. But over the years he found that the concept did not harmonize with what he observed on the landscape so he successively discarded parts of it until eventually it disappeared entirely from his use.[10]

After about 1950, by which time the first generation of American plant ecologists had largely lived out its allotted span, attention in the science broadened. Now ecologists considered their main objects of study to be *ecosystems*, interacting systems of plants, animals, microorganisms, and nonliving factors of the total surroundings. They now became concerned with aspects of the living world which were largely not considered by the first generation: food and predator-prey relations, cycling of nutrients and essential materials, the flow of energy on the whole earth and subunits of it, and the modeling of ecosystems using mathematics and computer simulation. Above all, they placed emphasis on quantitative techniques and explanations; these involved using counting, measuring, and statistical analysis to arrive at numerical, not verbal, descriptions of ecologically important things.[11]

As a member of the first generation, Wells did not participate in these new thrusts. But there were forecasts of them in his work. In one study pursued late in his career he made counts of plant occurrences and used primitive statistical procedures to analyze his results. He did seek to understand whole ecosystems even though he studied plant communities that were only parts of those systems. At the beginning, in studying the Big Savannah, he paid close attention to the fact that certain plants characteristically grew on anthills, as well as to possible effects of earthworms and crayfish on soil properties and through these on plants. The best example of his thinking in terms of dynamic systems is his explanation of how cyclic relations among plants, dry soil, and fire determine the occurrence of the longleaf pine, turkey oak, and wiregrass community in the sandhills of North Carolina. He first described these relations in 1931, calling them "the vicious circle." By this explanation, high temperatures and the coarse sandy soils result in plants with low water contents and many dry, tinderlike parts. The consequent frequent fires lead to open stands of the dominant plants, which are among the few able to survive such conditions. The nature of the vegetation in turn ensures continued high temperatures at ground level and maintains the soil in its sandy condition with little organic matter. This cycle is repeated over and over. The pioneer's emphasis on succession and the newer ecosystematic viewpoint are both clearly revealed by these words from one of Wells's technical publications: "Ecological studies have repeatedly proved that all animal and plant problems may only be fully understood when seen in the framework of the whole complex of interlocking factors and in the light of the recent story of succession."[12]

On that fateful spring day in 1920 when Wells first so abruptly encountered the Big Savannah, his eyes opened wide to the beauty of it, as would have perhaps the eyes of most. But his mind opened also, as he perceived in a flash that the Big Savannah embodied countless fascinating scientific problems. More than this, his whole life opened up in a way he had never imagined. For thirty-four years thereafter he was consumed by the wish to know why plants in nature grow where they do. That overwhelming need led him to contribute his thoughts, his efforts, and his professional career to important and pioneering work in the young science of plant ecology.

CHAPTER 2

Pioneer Ecologist

What an enjoyable week that was! I was delighted to find that Wells was a man of quite unusual breadth of interests and understanding. Our discussions were by no means limited to coast ecology; they included the ecology of bogs, savannas, grasslands, balds, the effects of fire and of primitive man, the origin of the Carolina Bays and human ecology. I discovered, also, that he possessed wide knowledge of literature, music and painting. In many of the subjects under discussion our individual approaches were from different experiences and knowledge gained in different countries, and we had, therefore, to enter the common ground of fundamental ecological concepts. I was most impressed at the clarity of Wells's thinking and his capacity both of analysis and of generalization. Whether we were watching the dance of soil particles in a sand rift or discussing ecological principles, I was at the feet of a teacher who freely shared with me the thoughts of a more mature and abler mind.—A. W. Bayer, 1962

ALTHOUGH HIS MENTAL horizons were not so restricted, throughout his professional career as a plant ecologist Wells carried out his technical researches entirely within the state of North Carolina. Almost immediately after his intellectual collision with the Big Savannah he began to survey the vegetation of the state. The wheels of scientific publication turn slowly, but by 1924 he was able to see in print his first sizable ecological report, a technical bulletin of the North Carolina Agricultural Experiment Station titled *Major Plant Communities of North Carolina*. In this synopsis of vegetation types he recognized seven communities that had been listed a few years earlier by the vegetation cataloger John William Harshberger: the sea oats dune community, the coastal salt marsh, the freshwater cattail marsh, the waterweed community of quiet waters, the gum-cypress swamp forest, the hardwood forest of the piedmont, and the spruce-fir forest of the high mountains. To these he added the old-field and roadside herb community, distributed throughout the state, and three communities of the coastal plain: the shrub bog or pocosin, the longleaf pine–turkey oak–wiregrass sand ridge community, and the grass-sedge bog or savannah that had first inspired his interest. In summarizing these communities Wells went beyond the simple descriptions of species occurrence provided by previous writers; he also included for each community a brief discussion of habitat factors, landscape features, and inferred successional relationships. For its time, this work was remarkable; in only a few years he had been able to observe the vegetation and draw general conclusions about its ecological relations which were surprisingly accurate.[1]

Having surveyed the vegetation, Wells embarked on more intensive studies of particular communities. The one he examined in detail first was, of course, the extraordinary Big Savannah. He made it the object of a two-year study in 1925 and 1926. In designing this project he planned what may have been one of the earliest American cooperative programs in ecological research. Five members of the North Carolina State College faculty were to pursue different aspects of the investigation: Carlos Frost Williams, the anatomy of the savannah plants; Ivan Vaughan Detweiler Shunk, microbes of the soil; Luther George Willis, chemistry of the soil; Louis Jerome Pessin, physical factors of the habitat; and Wells, communities and successional relationships. Alexander Campbell Martin, Donald Benton Anderson, and Larry Alston Whitford also participated in the work. The Carnegie Institution of Washington, in the person of the noted Frederic Clements, was to cooperate in the study. The purposes were two: to learn about environmental and plant factors responsible for the occurrence of the savannah community, and to use this basic knowledge, if possible, to devise practices for making such areas agriculturally productive.[2]

The group aspect of this study was not fully realized. Clements did offer some advice and suggestions, and the others did make some contributions, but the work was mainly executed by Wells and Shunk. This was their first joint effort, and it marked the beginning of a long association. For about fifteen years Shunk collaborated with Wells, until his health no longer permitted strenuous field activity. In their joint work Wells was definitely the master: his was the inspiration, the imagination, and the driving energy. As Wells himself later put it, Shunk's "well stored mind was a good one against which to test ideas." Technical results of the Big Savannah study were published in 1928, and that same year Wells amplified his interpretations of the whole range of coastal-plain communities in a significant publication in the scientific journal *Ecology*.[3]

The rest of Wells's principal ecological researches followed in an order apparently determined by his interests and opportunities. His investigations were, with one exception, concentrated in the eastern part of the state, mainly in the coastal plain. For the next detailed study he turned to the sandhills region; he and Shunk worked in the longleaf pine–turkey oak–wiregrass sand ridge community from 1927 to 1929, publishing their results in 1931. This work was punctuated in 1928 by a brief examination of the shrub-bog pocosin of the "open grounds" of Carteret County, results of which appeared in 1929. Then in 1930 and 1931, with his graduate student W. Melvin Crafton, he studied succession in old-field herb communities near Raleigh, a publication resulting in 1934. Two summers of teaching in the mountains quickened his interest in the high-mountain grassy bald community, which he studied in 1935 and 1936 and published on in 1936, 1937, and 1938. Beginning in 1937 he was back in the coastal plain, this time working in the maritime zone next to the ocean. Here he and Shunk discovered the effect of salt spray on the zonation of vegetation on and near the strand, and he described the live oak forest community on Smith Island; accounts of these works were published in 1937, 1938, and 1939. World War II brought restrictions on travel by automobile, and this and other limitations reduced Wells's research output.

On a field trip, B. W. Wells with his graduate student S. G. Boyce explaining his concept of "the vicious circle" in sandhills ecology, 18 April 1953. Courtesy of James W. Hardin.

But he did publish in 1942 a scholarly and authoritative review of scientific literature on the ecology of the southeastern coastal plain. When the conflict ended, he was able to conduct a detailed study of the large Holly Shelter pocosin in 1945 and to publish a report on it in 1946. During the final years of his active ecological career, from about 1949 to 1954, he devoted his attention to the problem of the Carolina Bays, publishing his principal report on that subject in 1953.[4]

Wells accomplished all of these studies in the first-generation manner. He observed the natural landscape directly, often at the expense of great physical exertion in remote areas; his ideas were individually conceived; he had the benefit of little financial support; and he worked largely in the summer because during the college years he taught a full schedule of classes and administered an academic department. He not only made unusually acute observations of vegetation and environmental factors, he also devised closely reasoned hypotheses of explanation which were capable of being tested. These things are what the muse of science demands of her servants, and Wells attended her faithfully. His significant contributions to the emerging science of plant ecology are evident when his work is viewed as a whole and not as separate studies. They lie in the following six general areas: clarifying the ecological role of pine communities in the coastal plain; deciphering the complex interaction of topography, soil, and fire in stabilizing coastal-plain communities; elucidating the successional sequence of old-field communities; proposing an intriguing explanation of the high-mountain grassy balds; discovering the important role of salt spray in coastal ecology; and stimulating ecological interest in the origin and history of the Carolina Bays.[5]

The question of the ecological status of pines in the southeastern coastal plain is a recurrent theme running through much of Wells's work. When Europeans first came to the region, they found that upland forests on the sandier soils were dominated by pines, especially longleaf pine. When the naval stores industry, sawmills, and other human activities removed almost all of those trees, their place was taken by other kinds of pine, in North Carolina mostly by the loblolly species. Early students of the vegetation looked simply at distribution and not at ecological factors so they regarded these pine communities as a major vegetation type distinct from the deciduous or hardwood forest. The latter, with its characteristic oaks, hickories, maples, and beeches, predominates on the best sites throughout the eastern United States. Until the 1920s most vegetation scientists considered the pine forest to be a climax community determined by climate and called it by some such name as "the Southeastern Evergreen Forest."[6]

Wells believed that plant communities should be interpreted *ecologically*, with reference to factors of their habitats and their successional relations. This viewpoint and his sharp observation caused him to conclude as early as 1924 that the pine communities, although long persistent, were not true climaxes. He thought that, depending on the site, some of them were subclimaxes, arrested developmental stages in succession to the hardwood forest; others were essentially permanent because of highly sterile soils. His thinking was clear: it was not ecologically sound to place the pine forests in a separate vegetation type merely because those trees have a broader tolerance of poor sites and sterile soils than do hardwoods. "Either a non-ecological definition . . . must be given," he wrote, "or if a truly ecological statement is made, the southeastern pine communities must be included in the . . . [hardwood] forest." A few earlier workers had suggested that the frequent fires that occurred in the coastal plain might be involved in the distribution of pines. As he studied the communities more and more, Wells became convinced that fire is an extremely important ecological factor in the Southeast. By 1931 he felt sure that the longleaf pine community was maintained by fire in a subclimax position in the eastern hardwood forest.[7]

Whether the primary factor was soil, site, or fire, or whether these forces operated in combination, Wells had recognized that the southeastern pine communities existed as they did because of nonclimatic circumstances in special localities of the hardwood forest region. Assessing the situation, he wrote in 1942, "Early vegetation maps and writings based upon the common observation of predominance of pine in the coastal plain gave the impression that pines were climax in this region. The 'southeastern evergreen forest' was made coordinate with the 'oak-hickory forest' of the piedmont region next to it. It is now generally accepted that the extensive upland pine forests of the coastal plain constitute fire subclimaxes." Vegetation can, of course, be classified differently from different points of view; Wells and most other ecologists of the time wanted ecological classifications to be based on ecological relations. His contribution was to point out clearly and emphatically that on such a basis the pine communities had to be considered as parts of the hardwood forest.[8]

Although there were a few dissenters, mostly persons who had never been to the Southeast to see with their own eyes, the subclimax concept was generally accepted among ecologists. But Wells's contribution in establishing a basis for that concept was not so generally recognized. Over the long period from 1927 to 1978 only a few ecological writings on the subject cited his works; most did not mention them at all. As recently as 1973 one writer described the idea as if it were so universal a truth as to require no credit to anyone: "The plain ecological fact, that southern pines are subclimax to hardwoods and require extraordinary measures to maintain them, has long been known and today is conceded by nearly everyone." Before the work of Wells, the ecological fact was not so plain, and since then his role in establishing it for nearly everyone has been largely unrecognized.[9]

Wells's successful attempt to explain the ecological position of the pine communities led him to broader generalizations that applied to other communities as well. When he first surveyed the North Carolina coastal plain, he found a remarkable array of vegetation types. Some of these were similar to those found in other regions, but some were unique. Examples of the latter include the longleaf pine–wiregrass savannah and the shrub bog or pocosin, the latter dominated by such shrubs as gallberry, leatherwood, and honeycup. He also found that succession was not simplifying the vegetational mosaic as much as might have been expected because some communities were apparently changing only very slowly if at all. Faced in the coastal plain with what he regarded as an essentially uniform climate, he invoked habitat factors to explain the stability of the persistent communities. In deciphering the role of the pines as determined by interactions of factors, he grew to understand that such interactions occur in other communities as well. Eventually he worked out a comprehensive interpretation of how topography, soil, and fire operating together stop the march of succession, the particular community that is thus stabilized being determined by exactly how those three factors interact on a particular site.[10]

At first, Wells centered his attention on two factors: topography, the configuration of the land; and soil texture, the coarseness or fineness, which depends on the size of most of its small particles. He decided that on a given site both of these factors affected the nature of the occupying plant community by influencing soil water. Topography exerted its influence by determining the distance down to the water table and how quickly and completely the water drained off. Soil texture affected the rate of drainage and to a lesser extent the supply of essential fertilizing minerals to the plants. He also recognized the importance of seasonal changes in the depth of the water table below ground level, a characteristically variable feature of the coastal plain. To express this factor he coined the still ecologically useful term *hydroperiod* to mean "the time during which the water table is at or near the surface of the soil." With these considerations in mind, he at first categorized several coastal-plain communities as subclimaxes made persistent by topographic or soil factors.[11]

But Wells soon realized that fire was a major force in the southeastern coastal plain, interacting with the other factors in complex ways. His concept of such interaction was well developed by the 1930s and clearly described by 1942. The topography of the land might vary from flat lowland to flat upland to rolling upland; different kinds of plant communities, such as forests dominated by different kinds of trees, would occur in these places. But for a given topography, the kind of community would also vary with the water table and the soil. So, for example, on the flat lowland the mostly flooded silt soil would have a gum-cypress swamp forest, whereas the wet but not flooded sandy soil would have a hardwood forest dominated by different trees. Or, on the rolling upland the sandy slopes would carry an oak-hickory forest, while on the very dry coarse-sand ridges only oaks would be present. Further, for a given topography, water table, and soil the kind of community would depend on how often fires occurred. Thus on the dry, sandy, upland ridges the oaks would be accompanied by pines if fires were frequent, and a wiregrass community with few or no trees would occur if fires were extremely frequent. On the wet, sandy, flat lowland the hardwood forest would be replaced by a shrub-bog pocosin if fires were frequent or by a pine-wiregrass savannah if they were extremely frequent. The participation of fire in this interaction meant, of course, that the occurrence of the plant communities could be affected by human activities because fire can be caused not only naturally but also by the intentional or inadvertent agency of mankind.[12]

At least in principle, these views of Wells's concerning factor interactions have been accepted by ecologists generally. In 1948, for example, Henry John Oosting, a plant ecologist at Duke University, embraced them in his college textbook: "The height of the water table during the wet season and the amount of fire in dry seasons are fundamental factors in determining the nature of the vegetation." And a book on plant geography published in 1964 contained a discussion of the interaction of these factors in the region. But though the ideas were apparently accepted, Wells's contribution in developing the understanding was not clearly recognized for a long time. Since 1977, however, a number of ecologists have credited him with suggesting such factor interactions. Indeed, some of their modern analyses may be regarded as more sophisticated elaborations of his original basic concepts.[13]

Another subject which Wells investigated extensively was secondary succession in old fields. When agricultural cropland is abandoned, natural vegetation begins to occupy it, and a succession of communities occurs. The cropland was initially created, of course, by clearing a previously vegetated site. When such land is later left alone, the succession that occurs thus takes place on bare sites caused by disturbance of the original plant cover. Such succession is called *secondary* to distinguish it from *primary* succession on natural bare areas such as rock outcrops. These two kinds of succession have quite different sequences of plant communities, even in the same region. The secondary succession in old fields typically involves initial pioneer stages of herbaceous plants ("weeds"), some sort of intermediate community, and later forest stages. This kind of sequence was known to practical humans like farmers and foresters and to early ecologists as well, but only in a broad, general way. In the Southeast, one worker in Florida had studied the phenomenon but had emphasized the later tree-dominated stages. Wells's attention was drawn to the problem when he first began his ecological investigations because at that time examples of abandoned farmland abounded in the Southeast. He was the first to examine the early stages closely.[14]

In his 1924 bulletin on North Carolina vegetation, Wells described what he called the old-field, meadow, and roadside community. With somewhat different plants in different localities this community occurred throughout the state on formerly cultivated or otherwise cleared upland. He also described a three-stage successional series, which he found to precede the establishment of pines in the piedmont and eastern regions: annual pioneers (especially crabgrass), tall weeds (horseweed, dog fennel, and aster), and broomsedge (a tall perennial grass). The broomsedge stage, which lasted for some years, was almost always followed by pine trees. After this initial account, which was brief but perceptive, he subsequently published more detailed descriptions of the successional relationships of the old-field communities. In the 1928 study of the Big Savannah he reported the same succession occurring on abandoned fields at the bog margin, provided they were kept free of fire. In another report in the same year he described a similar successional series for the coastal plain; this one involved pioneer (crabgrass), middle (aster, Queen Anne's lace, dog fennel, and goldenrod), and final (broomsedge) stages, the latter being followed by pine. In 1931, as part of the study of the vegetation of the sandhills, he reported for dry sites a succession of herb communities which differed somewhat from the usual sequence but was also followed by pines. In 1932 the communities of old fields and their successional relations filled an entire chapter of his semipopular book, *The Natural Gardens of North Carolina*. Finally came the thesis study of his graduate student, W. M. Crafton, submitted in 1932, which was based on a detailed examination of abandoned fields near Raleigh and led to a published version in 1934. This work confirmed the sequence of communities previously reported and offered in addition some causal explanations for that sequence in terms of light, tolerance of different species to dry soil conditions, and characteristics of seedling growth.[15]

These contributions by Wells to the study of old-field succession went largely unrecognized by most ecologists. Apparently they have seen his association with the problem almost entirely through the report of Crafton and Wells in 1934; his earlier works beginning in 1924 were mostly overlooked, ignored, or dismissed, but in any case unmentioned. The reason seems to be at least partly associated with other ecological events that took place in North Carolina quite apart from Wells. In 1932 a young second-generation plant ecologist, H. J. Oosting, joined the faculty of Duke University and began to build one of the most prominent and influential schools of American plant ecology. Oosting promptly initiated an organized program of long-term ecological investigations in the newly established Duke Forest. Because this facility included numerous tracts of abandoned farmland, it afforded a good opportunity for the study of old-field vegetational development. Oosting began work almost immediately, and over about four decades thereafter a long series of his graduate students produced a steady stream of sound scientific research on various aspects of the problem. Their reports largely ignored most of Wells's old-field work. Many of them, although not all, did cite the 1934 study by Crafton and Wells, but mostly the earliest works mentioned were publications in 1938 and 1942, which came from the Duke program. This lack of attention from an influential quarter may have set the tone among ecologists generally. In twenty-three reports on the subject published from 1938 to 1983, a total of thirty-three ecologists potentially could have listed 138 distinct references to the publications. There were, in fact, only 20 citations, of which 12 were to the report by Crafton and Wells and 2 were to *The Natural Gardens*. No one cited the original 1924 description for its discussion of the old-field communities. The currency which scientists covet most in payment for their labors is recognition; in this respect Wells is still owed. In 1977, however, the Crafton-Wells report was included in a historical collection of classic works on succession. Such recognition may lead to more general appreciation of what Wells actually contributed in this realm.[16]

A totally different ecological problem was attacked by Wells in the middle 1930s, that of the grassy balds of the southern Appalachian Mountains. These features, which are distinctly different from the so-called heath or shrub balds and should not be confused with them, occurred as small grassy patches scattered infrequently through the forests on the high ridges of the mountains. The community occupying them was usually dominated by perennial grasses, especially wild oatgrass, although sometimes sedges predominated; trees were absent, and with few exceptions shrubs occurred only on the borders at the edges of the forest. The forest community commonly occurred at even higher sites so these grassy openings were not to be explained by elevation, as are the alpine meadows of the high mountains in the West. The grassy balds posed two separate ecological problems: what caused them to occur in the first place, and what factors permitted their persistence, which seemed to contradict the rule of succession. Several earlier workers had noted their existence and offered some explanations for them. These explanations were unconvincing, however, and the reasons for the grassy balds were a mystery.[17]

Wells first approached the dual problems of origin and maintenance with a study of Andrews Bald, a seventy-five-acre patch on a south spur of Clingman's Dome, the highest peak in the Great Smoky Mountains National Park. He followed this work with a survey of twenty-two such balds in an area extending along the mountains from Roan Mountain through Haywood, Jackson, Swain, and Macon counties. This work led to his publication of two principal and three minor reports in the years 1936 to 1938. On the basis of his field observations he proposed two principal explanations: the grassy balds were created by prehistoric human disturbance, and they persisted because shrubs and trees of the surrounding forests were unable to invade the grass community once it was established.[18]

Wells's hypothesis of grassy-bald origin was based on close ecological reasoning. On the one hand he felt that large-scale general factors such as climate or soil could not explain the scattered and localized distribution of the balds. "Any natural forces operating singly or in combination which could bring about bald formation," he wrote, "have had sufficient time in the milleniums past to have reduced large sections of the mountain tops to the bald state. Yet the grass balds are but mere dots on the mountain landscape." On the other hand he concluded that small-scale local factors such as fire, ice storms, wind destruction, or pest attacks would soon have been followed by well-known successions, which would have restored the spruce-fir forest. But if neither large- nor small-scale environmental factors could account for the balds, the most likely alternative was their creation by human disturbance. This conclusion was initially inspired by the typically Wellsian observation that the grass community typical of the balds was the same as that occupying old Indian trails on many forested ridge tops. "In fact," he wrote, "many long trail sections are nothing but linear grass balds, showing in every respect the true bald character." Such trails were created, of course, by human disturbance. Consistent with this conclusion was the fact that some of the balds had sharp, angular boundaries, sometimes with right angles, a characteristic hard to attribute to natural factors.[19]

At the time it was thought that many of the balds which Wells studied existed before Europeans settled the region. Furthermore, the Cherokees who had inhabited the area when white settlers came were agricultural inhabitants of the lowlands. To them also the balds were a puzzle, explained in legend and myth as the work of various supernatural creatures: an immense hornet, a great lizard with glistening throat, a huge snake, a king of the rabbits, a race of spirit folk, and a giant with blazing head. Because the humans who created the balds apparently preceded the Europeans and were not the Cherokees, Wells concluded that they must have been an earlier group with a hunting culture. He made corroborating observations that supported this idea. He noted personally that ascending and descending the high Appalachian ridges on the same day left little time for activities such as hunting or ecological observation; thus hunters, if not ecologists, would have needed camp sites at high elevations. Most of the balds he studied were located on gentle south-facing slopes on high ridges, gaps, or knobs and were close to springs. Such places were just those that would have been chosen for camps. The few balds that did not fit these specifications could be interpreted as lookout sites, signal stations, or game lures, open areas into which turkeys and deer could have been enticed for shooting from the surrounding forest cover.[20]

Wells managed to get fourth-hand anecdotal oral testimony that one particular bald which was clearly undesirable for camping, that on Mount Sterling, was in fact used as a game lure. He also considered that it would have been well within early human means, even allowing for crude stone axes and other implements, to create balds of the sizes found by the first Europeans, especially with the aid of fire and with trampling of any secondary woody growth that might arise. And a decade later he noted that other workers had described in other parts of the world unforested sites that might have been initiated by ancient humans. Such areas were extensive heather lands in Denmark and eastern England, possibly created by Neanderthal man, and certain grassy regions in New Guinea, which might have had an aboriginal human origin. Still later, another worker compared the balds to certain grassy openings in the high-mountain forests of New Zealand. These sites were occupied by the same type of grass characteristic of the southern Appalachian balds and were known to have been made by the Maori people. An ecological term for a community persistently maintained by human disturbance is *disclimax*. Wells thought that the grassy balds and the other cases mentioned probably all belonged to one and the same category, and he invented and proposed for it the term *archeological disclimax*.[21]

Wells's second hypothesis was that once the grassy balds were created by human disturbance, they persisted because shrubs and trees were simply not able to invade the established grass community. This explanation was weakly based. It grew largely from the observation that when he examined the balds such woody plants were not present. He also thought he discerned a successional sequence that restored the grass community after drought or heavy grazing by cattle disrupted it. But Wells examined the balds at essentially only one point in time; he visited most of them only once in one of two seasons. Uncharacteristically, he drew his conclusion too hastily. Nevertheless, he became and remained thereafter convinced that the wild oatgrass could prevent the establishment of woody plants, even though he wrote that "only long-time quadrat studies can disclose the facts here."[22]

Wells's hypothesis that the grassy balds were made by prehistoric humans brought him considerable attention but little respect. Plant ecologists in general rejected it as not even a remote possibility. A few missed a crucial point by thinking that he meant the recent agricultural Cherokees; these workers wasted considerable effort in attacking an idea that lived only in their own imaginations. One of them was a graduate student of H. J. Oosting's at Duke University, who in 1941 first misrepresented Wells's hypothesis by describing it in terms of the tramping feet of modern Cherokees. He then proceeded to attack the straw man by collecting negative anecdotal information concerning those recent Indians, including an opinion from the "historian for the State of Tennessee" that the idea of Indian creation was absurd. Most ecologists were more scientific in their approach but refused to consider Wells's explanation seriously because there was no positive evidence in the form of artifacts for Indian occupancy of the balds, a telling point indeed. These ecologists continued to suggest that environmental forces such as localized climatic conditions were responsible but failed to counter convincingly the argument that if such factors were the cause, then the balds should be much more numerous and widespread than they were. Wells continued to stress the latter point in short publications in 1956 and 1961.[23]

Perhaps the overwhelmingly negative reaction that greeted Wells's view of Indian origin of the balds was owing to the mind-set of plant ecologists of the day. Most of them preferred to think of the world of vegetation abstractly, as it would be without human interference. So they felt comfortable only with explanations based on "natural" factors, which for them did not include humankind. But Wells's intellect operated freely across pigeonholes of intellectual convenience, and in this case his thinking had led him over disciplinary boundaries. His ecological reasoning had produced a conclusion that was anthropological and would therefore need an archaeological test in the form of diggings on the balds, something no ecologist was prepared to undertake. But in 1970 Philip Joel Gersmehl, a geographer neutral in ecological disputation, focused the lens of his specialty on the problem of the balds. On the one hand he pointed out that the so-called legends of the Cherokees were short-lived and were thus untrustworthy evidence that the balds antedated European settlement. On the other hand he did conclude that the grassy balds originated as cultural artifacts through grazing and fire, although probably not at the hands of Indians. In 1980 Garrett A. Smathers pointed out with some courage that Wells's hypothesis was intriguing because there is some evidence that although the ancient Indians of the mountains lived in valleys, they moved to summer camps at high elevations. This worker concluded that "it is not likely that the Wells hypothesis will find wide acceptance until well-disciplined archeological research is undertaken. The Wells hypothesis is tenable and it deserves to be tested. It is surprising that it has not been vigorously tested before."[24]

Wells's second hypothesis, that the grassy balds persisted because shrubs and trees could not successfully move into the grassy community, was apparently dead wrong. With the passage of time, human disturbance of many of the balds has lessened or been removed completely. And ecological studies conducted between 1953 and 1980 have revealed what Wells could not have seen, that with less or no disturbance most balds can be invaded and supplanted by the forest community. Ironically, although establishment of the fact that the balds persisted because of human agency destroys Wells's second hypothesis, it adds a piquant circumstantial note to the possibility that his idea of Indian origin was correct. If maintenance of the grassy balds required human activity, perhaps their creation did also. The question then is simply which humans.[25]

From the problem of the grassy balds, Wells next turned his attention to coastal vegetation. Woody plants, shrubs and trees, growing close to the ocean, display a marked modification of form, the growth of the plants being greatly suppressed on the seaward side. The characteristic resulting shape, in which the plants slope away from the sea, had universally been called the *wind form* by botanists because they thought it was caused by the drying effect of moving air. Wells himself had invoked this explanation in 1931 when describing damage to longleaf pine trees growing near the coast. Plants of similar shape occur on high mountains, and the cause was assumed to be the same in both cases. But in the spring of 1937 Wells and Shunk, working on the seaward side of the Cape Fear Peninsula below Wilmington, began a new ecological study by tagging young branches on some "wind-form" yaupon and wax myrtle shrubs. On returning to the same plants two weeks later, they found that severe injury had occurred to the new shoots on the seaward side, the most injured parts being those most exposed. When they checked the records of the United States Weather Bureau in Wilmington, they found that at one time during the two-week period a very strong southeast wind had blown in from the ocean for a period of nineteen hours. The weather data also showed that the sky had been cloudy during the time of the wind, and frequent spring rains had added much water to the soil in the area. Wells concluded that his tagged plants should not have experienced drying: plenty of water was available to them, and evaporation from them should not have been excessive under cloudy skies, despite the wind. Even in the face of accepted dogma, he was not afraid to go where his reason led him. He decided that the real culprit causing injury must be some other factor associated with the wind, not the wind itself.[26]

Back in the field, Wells and Shunk found a broad shrub zone, dominated by wax myrtle, which faced southeast directly toward the ocean. Although the location of this zone was such that all the shrubs had been exposed to the southeast wind, injury to the plants was again not uniform; consistently, the farther a shrub was from the sea the less injury it had. Wells realized that the harmful agent must decrease in intensity with distance from the ocean and concluded that it had to be salt spray, the tiny droplets of seawater picked up by the wind, carried along, and deposited on objects in the path. He and Shunk then made extensive observations on the relation of natural injury to distance from the sea; these not only confirmed the explanation but also revealed that the distribution of different plants in the maritime zone was related to their tolerance to the salt-spray factor. They also performed simple experiments in which protected plants were sprayed with seawater, salt solutions, or plain water. They transplanted wiregrass plants from an inland sandhill community and found that those placed on a dune top were killed within a month, whereas those set out in a less exposed position behind the dune showed little injury. And using loblolly pine as an indicator tree because of its sensitivity to the factor, they were able to estimate the width of the effective spray zone at points along the coast from Myrtle Beach, South Carolina, to Kitty Hawk; the width of the zone of injury varied from one-tenth to one and one-quarter mile. They reported these observations and conclusions, arrived at in the first-generation manner of simple direct observation, in several publications in 1937 and 1938.[27]

As a result of his observations on the effects of salt spray, Wells realized that this is the factor that stabilizes the live oak forest communities which persist on the coast. He described such a community as it existed on Smith Island. And shortly thereafter he concluded that not only the live oak forest but also the seaside shrub and dune-grass communities are maintained by salt spray of different intensities. It was this appreciation of the role of a nonclimatic factor in determining the ends of successional lines which solidified Wells's polyclimax view of vegetational relations. "Stabilized by different salt intensities," he wrote in 1942, "the grass, shrub, and live oak forest communities are all climax on a stable coast line. And as such they may be dealt with satisfactorily only by a polyclimax scheme of classification."[28]

When Wells and Shunk published their ideas on the salt-spray effect, the initial reaction of ecologists was disbelief. In May 1938 Wells talked about the work before the North Carolina Academy of Science. After listening to him, William Chambers Coker, a patriarchal botanist from the University of North Carolina, pontificated in substance, "That's a good story, Wells, but we all know it's the wind." H. J. Oosting from Duke University was another who did not believe Wells's explanation. As he admitted much later in print and privately, Oosting sent one of his graduate students to the coast in a deliberate effort to refute Wells's work, a procedure scientists often euphemistically call "testing the hypothesis." When he heard about Oosting's response, Wells was furious: he thought that the Duke ecologist was deliberately trying to make him appear ridiculous and regarded the attempt as an insult to his scientific integrity. But although the Duke student examined carefully a number of environmental factors in the maritime communities, he quickly found that only salt spray was correlated with plant distribution. Wells was right.[29]

Over the years, the feelings between Wells and Oosting, kindled by more than one disagreement over ecological matters, sometimes grew hot. When the Duke student reported on his salt-spray study to the North Carolina Academy of Science in April 1941, he should properly have mentioned the initial discovery of the phenomenon by Wells and Shunk. But he did not. Wells was present, and he reacted with characteristic temper. In the words of an eyewitness, "Wells got very hot, you know, and said, 'Dr. Oosting, I'd like you and your student to know that this idea that you worked on was originally my idea, and I published a paper on it. You seem to have confirmed my work!' And sat down." Wells's outburst apparently did not modify the thinking of Oosting, who was at least as stubborn as Wells. The account of the Duke study published in 1942 contains no reference to the earlier work of Wells and Shunk. It did cite Wells's 1939 description of the live oak salt-spray community, but in such an offhand way that an uninformed reader would probably not have perceived the correct relation. Some years later, after World War II, another of Oosting's students presented an ecological seminar at North Carolina State College. A listener recalled how Wells responded after the young man finished speaking: "'Now, ——, since you have laid yourself open to this by your pronouncements. . . .' Then he proceeded to tear a lot of it apart. And it was a pretty heated kind of thing." One of Wells's faculty colleagues, a man inclined to understatement and to gentle views of people, thought that "Wells and Oosting . . . were always good friends. They just differed about interpretations of things quite markedly." Another colleague, however, believed that Wells never lost his anger at Oosting. And a third thought that it was simply characteristic of Wells's interesting personality that he often spoke with great force: "He was always so keyed up, so enthusiastic, so vehement in his approach to everything that the sparks would just literally fly."[30]

Whatever the feelings between Wells and Oosting, in this case of the salt-spray effect Wells was not to be denied fairly quick recognition for his discovery. Working almost simultaneously with Wells and Shunk, a South African researcher found the same result in that country, and the several reports from Duke provided convincing supporting evidence. Any lingering doubts were removed in 1954 by the detailed field and laboratory studies of Stephen Gaddy Boyce, a graduate student of Wells's. Over subsequent years Wells and Shunk have sometimes not been credited fully for their initial discovery, most notably by Oosting in his textbooks of 1948 and 1956. Nevertheless, today they usually receive acknowledgment for their pioneering work.[31]

Late in his professional career Wells became interested in the riddle of the Carolina Bays. These are shallow, sharply outlined, oval basins of generally small size, which lie spattered by the thousands across the coastal plain of the Southeast from southern New Jersey to northern Florida. One of their remarkable characteristics is their orientation: their long dimensions are generally aligned in a northwest-southeast to north-south direction, depending on the geographic region. Until the advent of extensive aerial photography in the 1930s, the presence of these features in such great numbers was not realized. Shortly after their abundance and striking orientation became clear, several scientists proposed that they were the result of a meteorite shower in which vast numbers of those objects struck the earth in a relatively short time. A romantic idea, this hypothesis of the origin of the bays attracted a few supporters and still hangs about, especially in minds unwilling to play by the usual rules of science. Wells himself at first flirted with the meteoritic explanation, in 1938 referring to the "supposed meteoric origin" of White Lake and in 1942 calling it "the best substantiated theory." Later he embraced the idea wholeheartedly. But geologists in general have felt from the first that no positive evidence supports it; it also cannot account in any known ways for most of the characteristics of the features, including their elliptical shapes.[32]

His defense of the meteorite explanation of origin aside, while studying the bays Wells made some acute observations that led him to a valuable conclusion concerning certain coastal-plain lakes. This hypothesis still holds merit today, and how he formulated it is an excellent illustration of the way first-generation ecologists could shrewdly extract a maximum amount of explanation from a few careful observations. Some of the thousands of Carolina Bays have bases of clay underneath. But a great many of them are now occupied by shrub-bog pocosin vegetation underlain by peat. Peat consists of partly decomposed remains of dead plants that were formed in areas of poor water drainage such as marshes, bogs, or swamps, usually under acid conditions. Because of the factors required for peat development, it is generally thought that the bays having it contained at some early time shallow lakes that were filled in by deposition of the peat. In the very small number of bays that include open-water lakes at the present time, each lake occupies only a part of its bay basin. The few workers who had examined these lakes before Wells thought that they simply represented the final stages of the filling process, and he at first agreed with this concept.[33]

But when he examined closely five of the lakes in Bladen County, Wells showed what the practiced eye of a first-generation ecologist could discern when carefully focused. For he concluded in 1953 that these lakes were not filling in at all. On the contrary, they were enlarging; wind-driven wave action was actually increasing their size by eroding away peat at their margins. His thinking was based on six observations, any one of which by itself might have amounted to little, but all of which taken together in his interpretive mind led to a new conclusion. These six observations were simple ones. The bases of the shrubs next to the water of a lake and those some distance from its margin were exactly similar. In response to fire, new shoots arise from the bases of the shrubs; the number of new shoots was just as great at the lake margin as at a distance from it. All roots of the shrubs which extended out into the water were dead or dying. Shrub masses that broke off at the lake margin because their root connections rotted away did not sink but floated away and were even washed up on the sandy beaches. The peat margin at the edge of the lake did not slope out into the water but either showed definite undercutting or was nearly vertical. And cypress and gum trees located out in the water had exposed horizontal roots. As Wells saw, these facts in combination were evidence that the lakes were expanding in size, not contracting.[34]

Wells reasoned further that if the lakes were expanding, then they must be not the original lakes, which would have been filled with peat, but rather newer developments. He proposed that such a new lake could have originated by fire destroying the deposited peat if the peat had dried out during a long period when the water table was low. This combination of events would have been most likely at locations near deep-river drainage systems, and the Bladen lakes are all near the Cape Fear River. When the water table rose again later, the burned-out depression would have filled with water, creating a lake. Wind-driven shore erosion would have steadily enlarged it, and the present orientation and elliptical shape would have resulted from the directions of the prevailing winds in the area, southwest, south, and northeast.[35]

At first, Wells's hypothesis of lake development was rejected, even ridiculed. His ideas are still controversial and have not won general acceptance. But his assumed mechanism of the shaping and orientation of shallow lakes by wind-generated waves is in accord with recent views of geologists. A few later workers have accepted the idea that some lakes are enlarging by erosion of their margins. And some ecologists rather grudgingly now admit that the origin of such lakes through fire in a peat-filled basin is a viable hypothesis. One recent worker, unable to account for the creation of a Virginia lake in any other way, concluded that "the supposedly mythical idea of origin through a deep peat burn cannot be eliminated."[36]

The professional ecological works of B. W. Wells, carried out entirely in North Carolina, made him the outstanding pioneer plant ecologist in the state. Indeed, a look back from the vantage point of the present shows him to have stood foremost among his generation in the entire Southeast. Although he was not one of those most influential in determining the course of first-generation plant ecology, his careful observations and thoughtful explanations had significance far beyond the state. His studies of southeastern plant communities were an important contribution to the known vegetational inventory of ecology, and his deciphering of the complex interactions of plants and environments also added breadth to the interpretive principles available to ecologists. The number of his ecological research publications, nineteen principal ones and twenty-one of minor character, does not seem large when compared with the lists amassed by the few giants among his first-generation contemporaries. But when the circumstances are considered, the volume of his ecological output is outstanding: he produced it all after the age of thirty-five while spending most of his time teaching during every academic year and administering a college department that grew from two to thirty faculty members over the span of his active career.

Wells's work brought him international notice. Harry Godwin, an eminent British ecologist, wrote in 1962, "Although I have never had the pleasure of meeting Professor Wells, the distinction of his scientific research and that of his pupils led me at an early date to enter into scientific correspondence with him and to initiate an interchange of publications which continues to the present day. I have the highest regard for his scientific achievements." That same year, A. W. Bayer, the prominent South African who had discovered the salt-spray effect almost simultaneously with Wells and Shunk, commented about meeting Wells during a tour of the United States in 1950: "I did not meet another ecologist who impressed me as having as wide or as clear a vision as had Wells. Much of Wells's work was done during a period when ecological thinking was clouded by the monoclimax concept of the climax. The fact that Wells's conclusions are now accepted and so often quoted by those who are today engaged in rewriting and revising ecological concepts, is a final verdict on the correctness and clarity of his thinking. He must be permanently honoured as one who, in spite of the clouds, modestly and clearly showed the way to the light."[37]

In his own country, however, Wells received little recognition from his ecological peers, both during his lifetime and upon his death. He was an early member of the Ecological Society of America and active in it, but he never held one of its elective offices. He chaired paper-reading sessions at various national meetings of the society and for many years acted as the representative for North Carolina on its Committee on the Preservation of Natural Conditions. He served as a critical reviewer of technical manuscripts for *Ecology*, the scientific journal of the society, and was a member of its editorial board for two years. Over the years the Ecological Society publicly showed its respect for almost all members of the first generation of pioneering American ecologists; measures of such respect included in various cases dedication of issues of *Ecology*, portraits published in that journal, a medal, specific citations, resolutions of respect with accompanying biographies, or brief obituaries. Wells received no recognition whatsoever, not even a simple death notice. On the contrary, after Wells's death the *Bulletin of the Ecological Society of America* refused to publish even a brief obituary of him on the flimsy ground that it would not have the space to do so for all the ecologists who would die in the future. Yet only several months earlier, that bulletin had published a notice of similar length in honor of another pioneer ecologist who was almost exactly contemporary to Wells in life span.[38]

Reasons for this lack of national appreciation cannot be cited with certainty, but there are some possible contributing factors. The small southern technical college at which Wells worked had no reputation for scientific excellence during his career, and academics are often apt to prejudge someone's work by the standing of his or her institution. The number of his principal ecological publications was not overwhelmingly large, but they were widely available and should have been noticed even though most of them appeared in journals that were no higher than second rank. Unlike some of his influential contemporaries, Wells built no network of disciples, a source of eminence for many scientists. He worked alone or with one principal collaborator, Shunk; and because of the nature of his college he was mentor during his entire career to only nine graduate students, only one of whom continued as an ecological professional. Most important, perhaps, were the facts cited earlier: some of his generalizations were accepted without the clear assignment of credit to him, some of his publications were ignored, and in several cases his hypotheses were apparently dismissed out of hand.

More than this, Wells's status with his peers was not merely that of unrecognized worth. He was apparently regarded negatively, at least by some. Such a situation arises often in science but usually is kept from public view, spread by word of mouth, and almost impossible to document. When one young ecologist was about to take up an academic position in North Carolina, a senior man counseled him to be extremely wary of Wells and his bizarre ideas. In fact, the only idea of Wells's that might justly be described as bizarre was his unscientific espousal of the meteorite explanation of the Carolina Bays, a notion that was neither ecological nor of his making. The reasons for such a negative estimation of Wells can only be surmised; perhaps they were engendered by loose talk among persons who had never read his ideas firsthand or considered them carefully.[39]

It may be that Wells's personality generated some of the negative reaction to him and his work, for he could sometimes serve as his own enemy. He was vigorous and dynamic in all that he did, warm and outgoing to people and full of enthusiasm for the world of science and the intellect. But these positive qualities were sometimes expressed unfortunately. He was completely open-minded when he approached a new problem, but when he had worked out an explanation that satisfied his own head and heart, he clung to it fiercely, thereafter deaf to alternatives. Argumentative in debate and discussion, often given to showy rhetoric, he was occasionally guilty of glib dogmatism. He could not tolerate sloppy thinking and shoddy science, which abound in this imperfect world; he squelched such error wherever he found it. But sometimes he mistakenly imputed it to persons who disagreed with him for sound reasons. The response of other ecologists to these traits may have influenced the extent to which they accepted or did not accept his ideas.

But those ideas were incisive. Some of them were so right for their time that they slipped into the body of accepted explanation almost immediately. Others were dubious but so provocative that they were challenges stimulating their own rebuttal. Most of them were bold. Wells observed keenly and offered ingenious and stimulating explanations. Right or wrong, his work merits attention because "a scientist is to be judged by the questions he raises, not by the questions he answers." Wells asked powerful questions. Although in his lifetime he did not achieve great renown among American ecologists, recently some evidence of growing recognition has appeared.

In 1980 one researcher described him as "the leading student of the coastal plain vegetation." And in 1978, the year of his death, the Sixteenth International Phytogeographical Excursion by ecologists of the world took place in his beloved Southeast; the volumes that later reported on this event were dedicated to him. Perhaps these small signs portend that proper recognition will come.[40]

CHAPTER 3

Complete Botanist

From researches prosecuted and published, the following may be of especial interest: 1. Introduced concept of evolution into the study of insect galls. 2. Developed a satisfactory theory to explain the tissue changes involved in the self-splitting leaves of certain kelps. 3. Described numerous new species of insect galls. . . . 9. Worked out the geological history of the lower Cape Fear Peninsula, proving that the coquina rock was of Talbot age and not the more recent Pamlico.—Bertram Whittier Wells, about 1970

Wells's interest in plants was first sparked when he was a youth, and in the beginning he was fascinated by everything about them. As he progressed further in botany, he necessarily specialized, although throughout his life he maintained an unusually broad perspective on the subject. His first particular research specialty was cecidology, the study of plant galls. A cecidium or gall is a localized abnormal growth on a plant, which is produced in response to the presence of another living organism. Such growths may occur on any part of a plant but are especially common on leaves and twigs. The agents that cause galls to form include bacteria, fungi, roundworms, mites, and insects. Some galls are simple, consisting of rather unorganized swellings of parts of the plants. Other galls are complex and have a definite structural form that varies specifically with the kind of plant and the particular agent causing the gall. Such a complex gall is biologically remarkable in that the growth processes of the plant are somehow taken over and controlled by the gall-former, a totally different organism. Galls caused by insects are among the most common and most complex. For many of these, once a correlation has been made, the identity of the insect can be ascertained immediately from the form of the gall. Some insect galls, especially those occurring on the twigs of oak trees, are rich in chemical substances known as tannins. In the past this fact made insect galls economically important in leather tanning and in the manufacture of inks.

Wells first became interested in insect galls while he was an undergraduate student at the Ohio State University, probably at the instigation of Professor John Henry Schaffner. A few years later, he collaborated on the subject to a slight extent with Paul Bigelow Sears, another faculty member at Ohio State. After graduation, while teaching at the Connecticut Agricultural College in 1912 and 1913, he collected gall specimens in preparation for a program of research on the anatomy of abnormal plant parts. In the course of eleven months he gathered 204 different galls, 22 of which he reckoned were so far unreported in the United States. In 1914 he published an account of the new galls. In this paper he lamented that insect galls had been so little studied in this country; his report, he hoped, might "act as a stimulus to the collection of cecidia by showing the unworked condition of the field." He devoted almost all of his attention to describing the galls themselves, leaving to entomologists the task of identifying the responsible insects. At the time, it was the custom of some workers to give scientific names to unknown insects merely on the basis of the galls they produced, but Wells deplored this practice and refused to engage in it. "New names of gall producing forms," he wrote, "should appear only with adequate descriptions of the . . . [insects] concerned." From the first his interest in galls was on the side of the plants, and so he became one of the few Americans and few botanists, as contrasted with entomologists, to study them.[1]

Once he began, Wells pursued research on insect galls intensively and wholeheartedly. Over the years during which this interest was paramount, he produced a total of eleven technical research publications on the subject. Some of these were routine but sturdy descriptions of previously unreported galls. In addition to the 1914 report from Connecticut, he produced in 1915 a study of galls occurring on hickory trees, in 1921 an account of thirty galls from five states, and also in 1921 a description of a new species of oak gall and its insect maker; the latter was a joint effort with his colleague and friend, the entomologist Zeno Payne Metcalf. These works brought Wells notice among other scientists in the field, but only the study of the hickory galls engendered significant comment. Professor Alessandro Trotter of the Royal School of Viticulture and Enology in Avellino, Italy, was one of the leading European workers on galls. He was also founder and editor of the scientific journal *Marcellia*, which was devoted exclusively to the subject. Trotter reviewed Wells's paper on hickory galls favorably in 1915. It was, he commented in Italian, "an interesting descriptive summary" with "many clear figures." About seventy years later, another scientist, writing on the geography of gall insects, found that Wells's description of so many different galls on one small group of trees, the hickories, was the best available example to illustrate an important geographic point. This writer not only referred to the 1915 paper but also reproduced all thirty-four of the beautiful drawings Wells had made to illustrate his publication.[2]

The capstone of his descriptive work on galls was the thesis Wells presented in 1917 for his Ph.D. degree from the University of Chicago. In this massive treatise he systematically described 792 galls of the northeastern United States and eastern Canada caused by insects, mites, and roundworms. He grouped the galls by the kinds of plants on which they occurred; for each kind he provided a key to the gall species, about half of which were illustrated by his own drawings. He subsequently enlarged this work until it included more than 900

B. W. Wells, at right, in the botany laboratory at the Ohio State University, with his friends Lyonel King, at the microscope, and Bentley Fulton, about 1910. Courtesy of Maude Barnes Wells.

kinds of gall. When he began it, nothing like such an extensive treatment had been done before for North America, and had it been published, Wells's status as an authority on galls would have been permanently assured. But although a brief summary of it appeared in 1918, copies of which were privately distributed by the University of Chicago Libraries, he was never able to find a publisher for the entire work. The reason may be that Ephraim Porter Felt, an entomologist already known as an expert on galls, produced in 1918 a similar large treatment of American galls as a bulletin of the New York State Museum. This publication by Felt probably preempted the scientific territory. Perhaps in the vain hope that it might someday appear in print, Wells retained his own treatise throughout his active career. Finally, in 1962 he deposited the manuscript and all supporting gall specimens with the Division of Insects of the United States National Museum, a part of the Smithsonian Institution.[3]

Although such descriptive studies were sound contributions to science, Wells's most significant researches on galls were more than catalogs of species. His thesis for the M.A. degree at Ohio State, published in 1916, was important. If not the first, this study was certainly one of the earliest to compare the structures of almost all known galls that occurred on one kind of plant, the hackberry tree. It also related the anatomy of the galls to that of the normal tissues of the plant. In reporting on this study, Wells sounded themes that would recur in his work. He again took the position that insects should not be named scientifically only from their galls before any actual rearing experiments identified the responsible makers. "Only the paleo-entomologist," he wrote, "should have the privilege of dealing in fragments." He also emphasized that the adult insect and the plant gall are both the products of the developing immature insect larva. And he pointed out that insect galls are a perfect material with which to study the mechanism by which hereditary factors come to expression in actual characteristics. As soon as it appeared, this work was noted favorably by Trotter, who called it "an interesting anatomical study." In this country Mel T. Cook reviewed the work twice, dubbing it "very interesting," "well illustrated," and "very important."[4]

Wells continued his work on hackberry galls by studying the very early stages in their development, using specimens collected in Kansas and Texas; he published a report on this investigation in 1920. Trotter also reviewed this work favorably, giving it a lengthy summary. Wells's two publications on galls of hackberry have been referred to by numerous later investigators. Their continuing importance has been signaled by citations to them in publications by others in 1931, 1945, 1953, 1960, 1962, 1964, 1970, 1973, 1982, and 1983. As recently as 1984, a researcher pointed out that detailed anatomical studies on galls are scarce and called Wells's work one of the few complete examples.[5]

Wells's most provocative effort on plant galls was a proposed scheme of evolutionary relationships among them. In 1921 he first broached his ideas in a brief note, then followed with a lengthy discussion of them. He built on some earlier concepts of others but extended these into a comprehensive suggestion about how the galls might have arisen and changed in the long span of time over which organisms developed. He based his interpretation on the characteristics of the plant galls themselves, not on those of the causative animals. Nevertheless, he felt that it was initial changes in the heredity of the animals which led to the development of new gall forms; alterations in the animal reached expression in the plant, "an interesting situation to say the least." Specifically, he suggested that evolutionary changes in the animals led first to an inhibition of processes of plant development: the result was formation of the simple galls, which have indefinite structure. Then continued evolutionary changes in the animals resulted in the progressive production of the complex galls, which have such strikingly specific forms. In support of these views, he cited scientific evidence which he regarded as "overwhelming." Such evidence included the fact that the complex galls in developing seem to go through a phase akin to the unstructured simple galls, a kind of developmental repetition of their evolution.[6]

This publication on evolution evoked a swift and strong response from Alessandro Trotter. He chided Wells for not citing a complete list of earlier scientific reports in which the question of gall evolution had been mentioned, including especially two of his own; these constituted about thirty-five references dating from 1767 to 1920. Then in a long, carefully documented, and closely reasoned discussion he analyzed the ideas of both Wells and Alfred Charles Kinsey. In a publication appearing in 1920, the latter, more famous American student of galls and other matters had based his own evolutionary views on the primacy of the insect in determining the pattern of origin, although he also used gall characters and expressed surprise that no one had done this previously on any large scale. Trotter's discussion was thorough. It even reproduced exactly the complicated diagram Wells had drawn to illustrate his ideas. It also detailed Trotter's own opinions, which clashed with those of Wells. He felt that galls and their animal makers evolved independently, that the subject of gall evolution should first be approached by considering small homogeneous groups of galls rather than all of them at once, and that an evolutionary arrangement should be determined by the relationships of the animals, not the plants. This matter was evidently of great importance to Trotter. Not only was his own paper a large one, but also his disciple and biographer Mario di Stefano devoted an unusual amount of space to it in both a discussion of Trotter's life and an annotated list of his publications.[7]

Wells was always very proud of his paper on gall evolution, but apparently he never knew of this response his work evoked. In fact, he was evidently unaware that Trotter even existed. Some reasons for this suggest themselves: Trotter published in Italy and in Italian, a language Wells did not know, and Trotter's journal *Marcellia* was probably not available at that time in the meager library at North

Carolina State College. Wells's personality was certainly such that if he had known, he could not have resisted a rejoinder; this argument is supported by an actual event.

In 1923 M. T. Cook also reacted to the paper on evolution. In a discussion of the structure and evolution of galls, he pointed out that Wells, as had been properly acknowledged, built his arguments on a statement published by Cook twenty years earlier. He also claimed to have made some of the same points now detailed by Wells, and he characterized the latter's work as reaffirming his own earlier conclusions. Replying in the same year, Wells published a note by which "in the interest of gall science" he called attention to a number of "errors and deficiencies" in Cook's discussion. These involved what Wells termed a "highly misleading and incomplete" interpretation of certain ideas and a fundamental assertion which he labeled "entirely wrong." Following this response by Wells, the matter was apparently left to rest by both men. Aside from this skirmish with Cook and the considerable attention from Trotter, the publication on gall evolution has been little noticed over the years. A. C. Kinsey in his classic work on gall wasps listed a reference to it in 1930 but did not specifically discuss it. It was also cited, but without major attention, in works published in 1931, 1945, 1962, 1964, and 1965.[8]

The paper on evolution was essentially Wells's last on the subject of galls, for by that time he had become engrossed in ecological studies of the vegetation of North Carolina. His first graduate student, Alexander Campbell Martin, worked for the M.S. degree in 1924 on a problem involving galls, but after that the subject was no longer under Wells's active consideration. He did return to it, however, for his presidential address to the North Carolina Academy of Science in 1934. That discussion, entitled "Galls and 'Galls,'" was very general. Its main point was that while the simple indefinite galls might be considered disease responses of the plant, the complex galls should not because they have a genetically determined character. Alessandro Trotter, again probably unnoted by Wells, commented favorably on this paper, calling it an "interesting article of general character." Finally, in 1941 Wells wrote a review of the book *Plant Galls and Gall Makers* by E. P. Felt. This work, which dealt with about two thousand American galls, was a revision of that author's earlier bulletin, which had probably undercut Wells's own large treatise. In it Felt reprinted five drawings by Wells which had appeared in his research publications. Because Wells was still nursing his own unpublished manuscript, he can perhaps be forgiven for recommending Felt's book rather grudgingly as "about the only work" in which to find a summary of North American galls.[9]

Although Wells did not actively study plant galls after embracing ecology, he retained a lifelong passive fascination in the subject. This dual interest is evidenced by the way he chose to list his research specialties in successive editions of the directory *American Men of Science*: 1921, "plant pathological anatomy"; 1927 and 1933, "plant pathological anatomy; ecology"; 1938 and 1944, "ecology"; and 1949, 1955, and 1961, "insect galls; ecology of the southeastern United States." Even in retirement he still thought of the galls. In 1965, after learning about some scientific results obtained at the Rockefeller Institute, he wrote to suggest further lines of study which might be pursued. In response, the researcher wrote, "It is always gratifying to know, when one writes an article, that there are professional readers who find it stimulating, and particularly so when they are readers such as yourself, of the generation which created the biology my generation grew up on."[10]

Wells's most significant scientific contributions were in plant ecology and in the field of insect galls. But he also produced some other works not in either of these areas. His first piece of scientific investigation, performed when he was still an undergraduate student at Ohio State and published in 1910, was a microscopic study of certain marine algae. In this work he provided a description and an explanation of the events that occur when the bodies of those plants split as they develop. In 1945 this work was summarized in a classic treatise on algae, in 1951 it was cited in a section of a respected manual written by Wells's own former protégé, George Frederik Papenfuss, and as recently as 1969 it was mentioned as a basic work in its field.[11]

After this first paper, even though he subsequently worked intensively on insect galls, Wells remained interested in all aspects of plants. In 1919 he published a description of some abnormal flowers of wild onion. In these flowers, what should have been a stamen, one flower part, had developed instead into an entire little flower-within-a-flower. It took the sharp eye of a good field observer to notice these aberrant flowers, of which there were only two. But Wells spotted them in May 1918 growing in a mass of normal ones beside a railroad track in Arlington, Texas, where at the time he was teaching at Grubbs Vocational College. After moving to North Carolina, he reported on an unusual shoot of a princess-tree which his wife found in 1920. In one season this shoot had grown to a length of nineteen feet five inches. Wells expressed the belief that this was a world record for growth of a temperate-zone tree and titled his report "A Phenomenal Shoot." If it was a record, it did not stand for long. Eight weeks after Wells's note was published, there appeared in the same journal a description by William Frederick Prouty of a longer shoot of the same kind of tree. Prouty's shoot, in Chapel Hill, North Carolina, reached in one season a length of twenty-one feet six inches, and he headed his report "A More Phenomenal Shoot." Interestingly, Prouty, a geologist, was one of the few scientists who some years later supported the meteoritic theory of the origin of the Carolina Bays which Wells so vehemently espoused.[12]

During the course of his ecological study of sandhills vegetation in 1927, Wells discovered a previously unknown kind of pyxie or pyxie "moss," a small wildflower plant belonging to the same family as the more familiar galax. The common form of this plant, botanically named *Pyxidanthera barbulata*, occurs fairly frequently in the coastal plain, but the form he found was rare and localized in its occurrence. The new plant differed from its more common relative in leaf length, flower size, and some other ways. The difference that most impressed him was ecological: the habitat of the rare form was dry, whereas that of the common plant was significantly more moist. Wells thought he had found a distinct and separate species, but before claiming so publicly he made a special trip by automobile to the New York Botanical Garden to seek the advice of John Kunkel Small, a recognized authority on the plants of the Southeast.[13]

Some of the circumstances of that summer trip are worth noting because of what they reveal about Wells and about the time. He made the journey in his Chevrolet roadster, accompanied by Larry A. Whitford, a former student, who was then a junior faculty member at North Carolina State College. For financial reasons, the pair did not wish to stay in hotels and so took along camping equipment, including an army tent the poles of which were so long that they had to be carried slung beneath the axles of the car. They camped along the Potomac River outside Washington, D.C., in the mosquito-infested New Jersey marshes, and finally in the New York Tourist Camp in the northern Bronx. Present-day travelers to New York City would be astounded that while away from the campground they left their tent and belongings unsecured, without suffering any loss or damage. In addition to conducting his scientific business, Wells ensured that they saw the sights of the big city. They toured the botanical garden and the zoo; rode the subways; walked around town to the docks, the Battery, and the aquarium; viewed the unfinished Chrysler Building and ascended the Woolworth Building, at that time the tallest; enjoyed movies and vaudeville at the Roxy and Paramount theaters, with Will Rogers twirling a rope and telling stories; ate at Childs restaurants and marveled at the Automat; and saw two stage plays, a mystery called *The Spider* and *The Twelve Pound Look*, which starred Ethel Barrymore. Wells clearly made their stay include all the wonders small-town visitors were expected to experience.[14]

Reassured by Small and back in Raleigh, Wells published in 1929 a scientific description of his sandhills plant as a new species, *Pyxidanthera brevifolia*, that name being chosen because of the small leaves. This was the only new plant species Wells ever proposed, and afterward he was always both proud of his work and convinced that the plant was indeed distinct. Later specialists in the naming and classification of plants did not agree, however. In 1964 one worker reduced Wells's plant to the status of a variety of the common form, *Pyxidanthera barbulata* variety *brevifolia*. In 1968 another worker found that it could not be distinguished on the basis of the number of its chromosomes, a scientifically important characteristic. And in 1975 still others examined a wider range of populations of the plant and found a continuous variation both in plant characters and in soil moisture of the habitat; furthermore, all the populations would interbreed with each other. They decided that Wells had compared extreme forms of what was really the same plant and thus removed even the variety status from his discovery. These later conclusions upset Wells greatly: he died believing that his plant was truly a separate species.[15]

As an ecologist Wells was always concerned with and interested in geology, for the nature of the land underlying plant communities must be understood if their environmental relations are to be properly discerned. In the last years of his career he pursued some studies on geological matters at or near the coast of North Carolina. In 1940 he reported on a survey of peat in eastern Dare County, a piece of work made possible by a newly opened coastal road from Engelhard to Manns Harbor. He concluded that the peat had accumulated continuously because of a gradually rising water table associated with recent higher ocean levels. In 1941 he and Shunk examined layering in deposits of sand and peat on the lower Cape Fear Peninsula. In this study they found at one site still green leaves buried beneath a sixteen-inch layer of black peat; they concluded that a hurricane had been involved in the formation of this unusual deposit. In 1943 Wells reported on Blythe Bay, a feature southeast of Wilmington with characteristics that seemed to him to indicate a previously unrecognized change in sea level.[16]

These geological studies aroused in Wells an interest in the historical development of the North Carolina coastline. In 1944 he reported on his most extensive investigation of this kind, a detailed examination of the layers of sediments exposed at a sea cliff on the ocean side of the lower Cape Fear Peninsula. From this work he drew conclusions about the age and development of that peninsula in relation to ancient risings and fallings of the sea. In particular he decided that the Fort Fisher coquina, a deposit formed from shell pieces and coral fragments, was a shoal deposit built up seaward of an earlier and smaller Cape Fear; this contention was supported by later studies of other workers published in 1980. Finally, in 1947 he reported to the North Carolina Academy of Science on the curious occurrence of a small Carolina Bay within a larger one. Even after he retired, his interest in the geology of the coast continued: he became fascinated by the problem of the origin of the offshore islands or outer banks of North Carolina. In this matter he adopted an idea that was not new with him and has never received general acceptance among earth scientists. According to this notion, those narrow islands began as sand spits attached to the northern mainland and gradually built up to the south by wind and wave action. Furthermore, they are now being eroded away and moving inland, eventually either to disappear or to join the mainland. Wells never published on this subject, but as always when he embraced an idea he became an active and vociferous advocate of the hypothesis, delivering several seminars about it at North Carolina State College in the late 1950s.[17]

Wells's approach to the natural world was that of a basic scientist, eager for knowledge for its own sake. But he was also a true son of his age and a true exponent of the philosophy of the land-grant college, a complete believer in the betterment of mankind through science and technology. He was thus ever mindful that pure knowledge can generate technical applications. From the beginning in his ecological studies he was always watchful for instances in which his work might suggest new and profitable agricultural practices. One of the stated objectives of his first detailed ecological investigation, that of the Big Savannah, was "to practically apply this fundamental information to agriculture."[18]

Wells's service to the objective of application was more than merely that of the lip or the pen. He was an early advocate of judging the quality of possible farm and forest land by the natural vegetation occurring on it. In 1927 he wrote, "The type of native vegetation present . . . should prove of great value in passing judgment on the productivity of virgin land, or land that has been long abandoned." And in 1931 he asserted that "no advice concerning acreage purchases can equal that given by the native vegetation." In his studies of the coastal plain he observed that the best land for cultivation would be that carrying white oak trees. Yet many times he noted that such "oak flats" were left untouched while less desirable upland sites were planted to crops; he urged the reversal of this practice. In the sandhills he recommended that the driest sites, recognizable from the occurrence of wiregrass, be avoided for general agricultural purposes. For wet areas he made recommendations based on his observations of "the remarkable degree to which the native herbaceous vegetation responds to slight changes of drainage." With regard to forestry, he offered specific suggestions concerning how to improve practices when pines were desired and what sites to choose for best results.[19]

In two cases Wells drew agricultural conclusions that were sound for their time but were made incorrect much later by actual events and developments. He firmly believed, and consistently declared to the world, that upland grass-sedge bogs like the Big Savannah could never be cultivated successfully. Why he reached this conclusion, and why it subsequently became incorrect, is discussed later. Wells also concluded that the shrub-bog pocosin sites of the so-called open grounds could never be agriculturally useful because of the properties of the soil. Certain minerals were lacking in them, and there was present at shallow depth a distinct sticky layer of peat through which plant roots could not grow. The upper layer to which plants would be confined by this layer experienced water fluctuations from flooding to drought, which would make crop growth unsatisfactory. Wells's conclusion was based not only on reasoning but also on demonstrated fact: just before he first surveyed the open ground sites, a million-dollar agricultural development in the area had failed. But in recent times modern heavy equipment has made it possible to break up the impervious layer and thus, with the addition of appropriate mineral fertilizer, to put the land into profitable agricultural production.[20]

In one case Wells drew a practical conclusion from his basic ecological work which later developments proved to be eminently correct. This was his conviction that the commercial culture of blueberries in the coastal plain would be not only possible but profitable. At the time he began his study of the Big Savannah in 1925, there was no commercial blueberry production in North Carolina, although wild plants were picked over for local use. Wells saw the possibilities, and so in connection with his ecological work he set out blueberry plants in several different habitats in and around his research site. The plants he used were "the large hybrid varieties," probably some of the first ones to be released to the public. He reported the results in his 1928 publication on the grass-sedge bog: conditions in the bog were unfavorable for the blueberries, but they did well on the lower flat areas of the coarse sand ridges where wiregrass showed good growth. Wells was optimistic. "Areas where the permanent water table is from two to four feet beneath the surface," he wrote, "should prove very favorable for these plants." He also set out about one hundred cranberry plants in the bog itself. These did well over a short time, growing and setting fruit; however, commercial success with them seemed to him to be unlikely.[21]

When Wells turned in 1927 to his ecological study of sandhill communities, he was determined to continue the testing of blueberries. He obtained "plants of the recently developed giant blueberries" from "the Whitesbog Company, of New Jersey." He set these out in Pender County along a transect that included soil and drainage conditions varying from wet muck-bog to level wiregrass semibog. He reported the results in his 1931 publication on the vegetation of the coarser sands; plants in the wet regimes failed to do well, but those in the moist wiregrass areas made excellent growth and bore abundant fruit. His conclusion and recommendation was clear and emphatic: "In the southeastern coastal counties there are thousands of acres of wiregrass semibog which from our tests should prove excellent for blueberry culture. . . . It is to be hoped a thorough trial of blueberries on such areas may be made by those farmers who have this type of land in their holdings." This research indeed pointed the way toward what is today an important commercial use of the land in North Carolina.[22]

In later years Wells often expressed the conviction that he had contributed significantly to the establishment of blueberry culture in North Carolina. Certainly he had the vision to see the unrealized potential of an economically important agricultural resource. But historical accounts of the development of commercial blueberry production in the state have made no mention of his work. The cultivated blueberry industry in the United States was based on the contributions of Frederick Vernon Coville of the United States Department of Agriculture, who early in the twentieth century selected wild blueberry plants and hybridized them to develop improved characteristics. Coville enjoyed the close cooperation of Elizabeth C. White of New Lisbon, New Jersey, a commercial grower of cranberries and blueberries on her Whitesbog Farm. Using the improved varieties, the first of which were released in 1920, the blueberry industry spread from New Jersey, becoming established in North Carolina in about 1930. The North Carolina Agricultural Experiment Station in its annual reports was silent on the subject of blueberries until 1938, when a breeding program led by E. B. Morrow was established to make use of plants left by Coville. The latter had begun in 1928 a cooperative program with the North Carolina Department of Agriculture in which test plantings had been set out on the Coastal Plain Station at Willard and on private farms. This cooperation continued during the 1930s and attracted both commercial interests from New Jersey and local growers. By 1940 the state Department of Agriculture considered blueberry production to be "thoroughly established as a major fruit industry of the section."[23]

Wells's test plantings of blueberries in 1926 and 1927 were clearly among the earliest, if not the first, in North Carolina. The plants he used were hybrids, some of them from Elizabeth White's Whitesbog operation in New Jersey, so he was evidently in communication with those in Coville's program. His plantings were accessory to ecological studies that were officially part of the work of the state agricultural experiment station, one of which was formally published as a bulletin of that organization. Thus what he did and what he recommended must have been known to agricultural leaders in the state and elsewhere. Although the true impact of his work is hard to assess, it seems likely that he did indeed contribute significantly to the early development of the North Carolina blueberry industry. If so, his contribution has so far gone essentially unrecognized.[24]

PART 2

Champion of Nature

CHAPTER 4

Conservationist

As one who in 1922 hiked up the old trail to Indian Gap (New Found Gap had not yet been found) and the crest trail to Clingman's Dome, I am in a position to know much about the scientific and scenic value of the high Smokies. I am going to be blunt. To put any more roads across this surviving wilderness will be a social crime! It will cheat the coming generations of sensitive lovers of the natural world from experiencing the undisturbed grandeur of our longest and highest eastern mountain range. The people who are really interested in and appreciative of our mountain preserves are always willing and glad to hike many miles to enjoy them. To open up this proposed road will be a start on the way of developing another Maggie Valley. Let me repeat. It will be a social crime to initiate any further inroads on this remarkable wilderness area.—Bertram Whittier Wells, 1966

B. W. Wells was the equal of any of his contemporaries in his belief in the technological progress of mankind. "I have always been one of those chaps that enjoys trying to do things that are new, somewhat different," he said at the age of ninety-three. But from childhood he adored the natural outdoors and fervently wished that as much of it as possible could be kept. In maturity, when he found the remarkable array of vegetation contained in North Carolina, he became its active champion. His outlook was twofold: the beauty of nature should be conserved, and people should learn, or not forget, how to experience it. Thus for the rest of his life he strongly advocated the preservation of key natural areas for all to enjoy, and he vigorously worked to publicize them so that all might want such enjoyment.[1]

An illustration of Wells's outlook in this respect is his reaction to the case of the longleaf pine forests of the coastal plain. Early colonial settlers and naturalists found and described majestic stands of this impressive tree, which on the uplands formed extensive unbroken canopies over a deep carpet of fallen needles. Long before Wells's time they were gone, and he was shocked and dismayed at what he found. "Not a part of this great natural wonder, worthy of the name forest, remains intact within the state's borders," he wrote. It had been, he lamented, "rooted out by hogs, mutilated to death by turpentining, cut down in lumbering, burned up through negligence." He thundered indignantly that "the complete destruction of this forest constitutes one of the major social crimes of American history."[2]

Wells's spirit was always one for preserving nature, but the major conservation effort in which he participated most directly and vigorously failed. This was a campaign to set aside and preserve his beloved Big Savannah for public enjoyment. Such preservation would have involved certain problems, however, in addition to the usual one of financing. Understanding those problems requires noting the special combination of ecological circumstances, first elucidated by Wells himself in his scientific study, which permits such a savannah to exist in the first place. These conditions are mainly a high water table underlying the land, an impervious soil that results in poor drainage of water, and—especially important—frequent fires, at least annually. Also contributing are a high soil acidity and almost no available soil nitrogen. Such factors taken together are wholly unfavorable to trees and other woody plants, which therefore cannot take over the landscape through the process of succession. This combination of conditions existed naturally in the Big Savannah and would have had to be perpetuated if it were to have been preserved. Elimination of those conditions, especially through increased drainage and suppression of fire, would destroy it forever.[3]

The opening battle of the campaign for preservation was an effort to induce the state of North Carolina to acquire and maintain the Big Savannah as a park and wildflower preserve. In the years 1935 to 1937 Wells stimulated the Garden Club of North Carolina to work diligently toward that end. In 1935 Mrs. B. E. Jordan, conservation chair of the club, petitioned the governor to set up a panel to study the proposal, and a Mr. Bland, acting for the club, secured options on most of the land. Wells and Mrs. Jordan also appealed personally to a finance committee of the state legislature to purchase the area, without success. The effort was continued the following year when Beulah Averiet Parker, the new conservation chair, organized a campaign of letter writing by club members. Also in 1936 the matter was taken up by the legislation committee of the club, which pursued an alternate course by approaching the federal government through its Resettlement Administration. This depression-spawned agency, the forerunner of the Farm Security Administration, had as one of its activities the purchase of submarginal land of no agricultural value, a category into which the Big Savannah certainly fell at the time. But the legislation committee learned from James M. Gray, associate regional director of the agency, that all available funds had already been allocated. The committee then submitted a resolution petitioning Gray to give the project first consideration should additional money be granted to the state.[4]

The garden club continued its effort in 1937. Parker reported as conservation chair that she had held several interviews with officials on the matter and expressed the hope that her successor could "put this over." A list of twelve suggestions sent to local clubs included one that they use their influence to secure the acquisition of the tract as a state park. The legislation committee continued to deal with the Resettlement Administration but learned that appropriations originally allotted to the southern region had been diverted to drought-stricken states of the West; these were the days of the dust bowl. So the committee turned back to the state government and attempted to have

B. W. Wells, in hat, on the last remnant of his beloved Big Savannah, with ecologist A. W. Cooper, hidden, parks representative Loddie Bryan, and student Charles DePoe, 8 May 1959. Courtesy of Arthur W. Cooper.

the legislature consider a bill authorizing the purchase of the area as a state park. But J. S. Holmes of the state Forestry Service objected because maintenance of the wildflowers on the tract would require frequent burning-over. For this reason, as well as others, the state Park Commission would not recommend acquisition to the legislature. Faced with such a negative position on the part of a key state agency, the garden club gave up the fight.[5]

Nevertheless, *North Carolina*, the Federal Writers' Project guide to the state published in 1939, included the Big Savannah as a stop on one of its recommended tours. The description called it "an area noted for the variety of its wild flowers and shrubs" and emphasized the curiosity of the Venus flytrap plant; it also called attention to swamp forests and Spanish moss. Wells contributed to this book a chapter on the natural setting of the state, so it is not surprising that the area was given notice. Curiously, even though the tract had no official status, the transportation map that accompanied the guide showed it in the green color used for parks and designated it by the Wellsian name of the "Big Savanna Natural Garden," as if it were already set aside. This misleading presentation may have helped foster the misunderstanding in conservation circles which later contributed to the disastrous final result. The revised edition of the book, published in 1955 as *The North Carolina Guide*, reprinted the reference to the Big Savannah without change.[6]

Although the effort of the garden club failed, the Big Savannah continued in existence. Wells still wanted everyone to experience the beauty that moved him so deeply; therefore he continued to push the idea of a park in his many speeches and field excursions. For example, in addition to his annual college class field trips, in 1953 he conducted a group of botanists there. For another example, in 1951 he spoke to the Cape Fear Garden Club of "the tourist value to Wilmington and the surrounding area . . . if it could be taken over and controlled by the State as a public resource." The elderly owner of the land also recognized its unique beauty and potential value as a wildflower sanctuary so at some point he offered it to that very Cape Fear Garden Club at a cheap price, less than two dollars per acre. But the club declined the offer in the belief that the land could not be used agriculturally and would thus remain in a natural state with no effort on its part. Looking back in 1959, a noted conservationist judged that "the failure of the Cape Fear Garden Club to act in this matter is almost incredible." Ironically, Wells himself was probably responsible for the attitude of the club. He was totally convinced that the tract could never be farmed because existing machinery was incapable of altering its drainage pattern, and he never hesitated to say so forcefully.[7]

Knowledge of the Big Savannah was spreading outside the state, partly because of the early technical bulletin by Wells and Shunk, but mostly because of Wells's book, *The Natural Gardens of North Carolina*. Such awareness was helped by the writer on subjects of nature, Edwin Way Teale. Teale visited it in 1947 and in 1951 described its beauty in his book *North with the Spring*. In 1949, before that book appeared, Teale called the Big Savannah to the attention of Richard Hooper Pough, suggesting that it was probably available for a state or federal park. Pough, an authority on birds, moved in national conservation circles. He was then curator of conservation of the American Museum of Natural History in New York; later he was a trustee of the American Scenic and Historic Preservation Society, president of the Natural Area Council, and president of the national Nature Conservancy. His interest aroused, Pough immediately sought information from contacts in North Carolina. Sadly, at this point a misunderstanding arose which had severe consequences. Somehow the last likely chance to save the Big Savannah was missed. Somehow Pough was given the impression by several persons that it was already a preserved area. It was therefore erroneously listed as a state park in the Nature Conservancy's preliminary inventory of natural areas, and Pough turned his attention elsewhere.[8]

By 1957, however, Pough made contact with Wells and learned from him the story of earlier efforts to save the Big Savannah. He apparently also learned that it was in fact not under protection, for he asked Wells in January 1958 exactly where it was located, whether acquisition of only a part of the tract would suffice, and whether a proper controlled-burning regime was practical. He wanted to move on the matter. "I'm afraid if we don't do something," he wrote, "we will suddenly wake up to discover that industry has found some use for it or some other disaster has overtaken it." How prophetic were his words. For whatever reason, Pough did not visit the site until March 1959. What he found there horrified him.[9]

Only about one-fifth of the Big Savannah, about 250 to 300 acres, remained. The owner of the land had died, and his heirs had sold it. When Pough arrived, most of it had successfully been put into the cultivation of truck crops. This conversion had required a complete and permanent alteration of the water-drainage pattern. An enterprising farmer from Ohio had accomplished this task with modern heavy equipment. Using a drag line, he established three-foot ditches; he plowed the soil deeply, turning it up into high ridges; and he limed and fertilized the soil heavily. Surface water thus ran off quickly into the ditches, and the crops thrived on the ridges. Fire, of course, had also been suppressed. Pough was beside himself. "I am appalled," he wrote, "to see it more than half destroyed. . . . Several people in the past have told me it was perfectly safe. What went wrong? If the State Legislature wouldn't put up funds to buy it, why didn't the Garden Club of North Carolina raise the funds to purchase it? . . . We just can't afford to let things like this go 'down the drain' without trying to preserve them."[10]

Faced with fact, Pough first planned to try to save the remaining fragment. "I am determined," he wrote, "to do what I can to help save what is left." He got a tentative commitment for a sale and immediately tried to stir into action both the Garden Club of North Carolina and the superintendent of state parks. The latter quickly sent representatives to examine the situation. They and Wells accompanied an ecology class from North Carolina State College to the site on 8 May 1959. When Wells saw for himself what had happened, his reaction was characteristically mixed. The conservationist in him deplored deeply the loss of his favorite spot of natural beauty. His progressive side was impressed and fascinated that technology and ingenuity had brought about the change he had thought impossible. On the basis of this inspection, preservation of the remnant was deemed impractical. It was left to its fate. By 1960 crops were established on the whole site. When Wells visited it again in May 1967, half of it, eight hundred acres, was one large cornfield and the rest had recently been plowed. In December 1959 Pough was beginning efforts to secure "the next best area that exists." But next best could never be the same. The farmer who initiated this complete transformation of an ecological marvel sincerely saw his action as progressive, not destructive. As Pough wrote, he believed that he was "doing something very laudable by putting into production land everyone thought was worthless." In fact, not everyone thought the land on which Wells's most beautiful natural garden stood was worthless. But those who did not failed to act decisively. The Big Savannah was gone.[11]

Wells also participated actively in another unsuccessful effort by the Garden Club of North Carolina, this one an attempt to have the Venus flytrap declared the official state flower. At the time North Carolina had no such flower, although the ox-eye daisy had often been given the title popularly. This plant, however, also known as the field daisy, white daisy, marguerite, or whiteweed, was not a native one, having been introduced and naturalized from Europe. To Wells the idea that such a plant might be the state flower was repugnant. In 1932 he called it in his book "this foreign weed" and asserted that "when North Carolina chooses a 'state flower' it will be a native North Carolina plant." Perhaps stimulated by Wells, the garden club rallied to the cause of certification of a native plant and in 1933 asked for nominations from its member clubs. The three most popular suggestions received were the dogwood, the azalea, and the Venus flytrap. Before choosing a candidate, the club asked for opinions from two botanists, Wells and W. C. Coker of the University of North Carolina at Chapel Hill. Wells responded with a vigorous letter in which he strongly urged the naming of the flytrap. It should be chosen "primarily for its scientific interest in addition to its beauty," he wrote, and "would reflect favorably on the intelligence of any legislature which made such a decision." Wells's appeal to scientific interest and legislative intelligence was probably not the best tack to take; nevertheless, the garden club made the Venus flytrap its official floral emblem, adopting it at the annual meeting in 1935. In the 1936 *Yearbook* of the club Wells continued the campaign with an article entitled, with characteristic hyperbole, "The Plant That Has Made North Carolina Famous." Eventually, however, the flytrap lost out in the state legislature to the more familiar and widely distributed dogwood.[12]

As a professional ecologist, Wells was a member of the Ecological Society of America. As a conservationist, he participated in one of the first activities of that society after it was founded, the Committee on the Preservation of Natural Conditions. This committee, established in 1917, had a twofold purpose: to compile an inventory of all natural areas in which "original" or essentially undisturbed conditions existed, and to work for the preservation of those that were not yet protected. The committee aimed to have a representative from each state or province in North America; from the beginning Wells and Z. P. Metcalf were the members from North Carolina. In 1926 an early fruition of the cataloging function of the committee appeared in the form of the comprehensive book *Naturalist's Guide to the Americas*; Metcalf and Wells contributed the section of the book which dealt with North Carolina. Wells was active on this committee until at least 1942. And from 1937 he served on one of its "special subcommittees to urge the preservation of small fragments of original biotic communities"; he was chair for the live oak and palmetto forest of Smith Island.[13]

Wells was one of the early members of the North Carolina Wild Flower Preservation Society, joining that group by invitation in 1951, the year of its founding; he continued active membership in it for the rest of his life. For many years he acted as a formal consultant to the society and contributed to it in various ways; one of these was by joining in its conservation efforts. For example, in 1956 the society, led by Lionel Melvin, mounted an effort to preserve the home of the sandhills pyxie, which Wells had described as a new species in 1929. Although very abundant at one location in Harnett County, this plant was known at the time only from that place, and the society felt that it deserved protection. Wells was a member of the committee that investigated the possible acquisition by the society of the tract of land on which the plant occurred. The desire to secure this site led the society to incorporate; by such a means ownership could be accepted from the lumber company that owned the land. This activity progressed as far as the society's securing of both a tentative agreement from the company to transfer the title and a promise from the state forester to protect the land once the deed was obtained.[14]

Wells also assisted the society in its conservation efforts in other ways. In 1956 Hollis J. Rogers led the organization in an effort to compile a list of wildflowers of North Carolina which needed protection. O. M. Freeman submitted a list for the mountains and Henry Roland Totten one for the piedmont; the list for the coastal plain was compiled by Wells. Then ten years later he joined with the society and many others in an outcry against the proposed construction of a highway through Great Smoky Mountains National Park from Townsend, Tennessee, to Bryson City, North Carolina. Wells's letter of protest was typical of him. First, he outlined the basis of his claim to expertise in the matter: he had helped influence the state legislature to provide money to purchase the park, and he had personally studied the area and written about it for the public in his book, *The Natural Gardens of North Carolina*. Then he unleashed his indignation in double-barreled bombast. The construction of the highway, he insisted, would be a destructive inroad on the grandeur of a magnificent area and nothing less than a social crime.[15]

Wells also wrote many letters in connection with other conservation issues. He served the Ecological Society of America as a representative interested in the protection of the live oak and palmetto forest of Smith Island. Therefore, when a movement arose in the 1960s to preserve that site from commercial development, he added his opinion to those of numerous other individuals and organizations. Testifying as a scientist, he wrote that the island was "an unsuitable location for the development of a resort." He based his position on five points, and as always his reasoning was ecological. The pervasive presence of salt spray would make gardens and other such plantings out of the question and would prevent the development of trees tall enough to provide shade. Leveling the ridges of the surface for housing would also cause destruction of the sheltering live oaks. Bathing would be unsafe because of strong currents on one shore and shifting shoal sands on the other. The tidal range of five feet and the swirling waters during tide changes would make the docking of small boats a difficult problem. And finally, the known rate of shore erosion on one side, one-quarter mile in the last hundred years, would endanger any permanent development. "It should be preserved just as it is," he concluded, "an amazing bit of shore line, little changed in historical time." Two years later, when it appeared that the effort to keep the island natural was doomed to fail, he maintained his ecological perspective: "I am convinced that attempts to make a resort there will fail and the Island will literally preserve itself. But don't try to put this over with . . . [someone] who believes money can do anything." Development of the site has actually proceeded in a manner different from that envisioned when Wells wrote; it remains to be seen how accurate his predictions were.[16]

Wells continued to write letters in support of conservation causes for almost all his life. He never lost the sense of wonder which the varied vegetation of North Carolina aroused in him when he first arrived in the state, and he went on trying to impress others with the need to preserve the best of it. As late as 1972, for example, at the age of eighty-eight, he addressed a newspaper to plead for conservation of natural areas before they became lost: "Large parts of them," he wrote, "should be purchased at low cost and made into parks, leaving the vegetation exactly as it is now. . . . North Carolina is the one Eastern State which still at low cost has available the greatest diversity and size of our wilderness areas."[17]

In his time Wells was not the only voice for conservation of natural areas in North Carolina, but he was one of few. Since his time the number has grown and today exercises greater influence and brings about more action than ever before. Yet the spirit of Wells is still felt. The program of the North Carolina Nature Conservancy to acquire and preserve choice examples of coastal-plain vegetation, such as clay-based Carolina Bays, would meet with his vigorous approval. And that organization is conscious of his pioneering guidance and inspiration: a recent publication proudly asserted, "Dr. Wells no doubt would be impressed today with the North Carolina Nature Conservancy's latest project located only a short distance from his beloved Burgaw bog." The Big Savannah is indeed gone, but Wells succeeded in infecting others with his love for such places, and sites preserved by them will now remain symbolic of that love.[18]

CHAPTER 5

Voice for Nature

It was like having the scales lifted from eyes long dimmed by defective sight to make an excursion the past week-end through the Coastal Plain of North Carolina . . . with a nature study party, directed by Dr. B. W. Wells, professor of botany at State College. Instead of marking the progress of the trip by sign-boards . . . that clutter up the scenery of the road, the trip became a series of ever-changing plant communities and regions of progressive geological interest. It was like seeing a drop of stagnant water spring to life with myriad protozoa under the microscope to note the trees and flowers take on individuality and variety in their own plant communities, in what had heretofore been simply stretches of woods and fields with little to distinguish them from each other.—Susan Franks Iden, 1926

Wells fought directly for conservation of natural areas, but he was most effective as a popularizer of what needed to be conserved. He loved the plants and vegetation of North Carolina, and he strove mightily to help others feel as he did. His views on conservation were not elitist; he believed strongly that people in general should learn to respect and value nature. Throughout his life he emphasized the idea that widespread popular appreciation was the principal purpose of preservation; as late as 1955 it was reported that he "especially stressed the necessity for educating the public to appreciate nature, saying that there is very little need for preserving plant life if there is no one to appreciate it." In this spirit, throughout his many years he carried his love and his science, his concern and his enthusiasm, directly to people through face-to-face meetings in lecture halls and on field trips.[1]

In the beginning, as he became more familiar with the vegetation of North Carolina through his technical ecological studies, Wells became more ready and willing to talk about it publicly. He acquired for use in his scientific work a four-by-five-inch Graflex single-lens reflex camera; with this device, huge and unwieldy by present standards, he became a skilled photographer and made many pictures of plants, flowers, and habitats. He thus assembled a large set of the black-and-white glass lantern slides characteristic of the time; these he tinted himself, using transparent watercolors. With such pictures at hand, he organized his thoughts into a lecture titled "The Patch Work of North Carolina's Great Green Quilt." Through the Extension Division of North Carolina State College he announced in 1925 his availability to speak on this subject. The published description of this talk carries traces of Wells's style of expression. It explains that the talk was organized around a botanical tour from the mountains to the sea and proclaims, "This lecture is illustrated by 120 lantern slides (can be shown in any room having electric wiring) which show in a striking fashion how the native vegetation of the state indicates the remarkable diversity of soil and climate within the state's area."[2]

At this time Wells was also prepared to deliver four other lectures through the Extension Division of the college. These were "A Biological Marvel: The Insect Gall," a description of "one of the most amazing and unique situations in the living world"; "The Age of Science," a survey of recent advances; "The Most Remarkable Plant Community in North Carolina: The Big Savannah," a discussion of habitat and plants, including the Venus flytrap; and "The Wild Flowers of North Carolina," with projected illustrations. His talks before school groups especially were so numerous over the years that available records would hardly begin to document them. One account noted, for example, that his speech "The Wildflowers of North Carolina" was given to the Parent-Teacher Association of the Dillworth School in Charlotte.[3]

Wells's public speeches were what first brought him to the notice of the Garden Club of North Carolina, and for at least twenty years he was in demand as a speaker to that group. In 1931 he delivered to the annual meeting of the club a wildflower talk, "The Natural Gardens of North Carolina," which spurred interest in his projected book on the subject and eventually gave it its title. In October 1935 he gave a lecture called "The ABC's of Botany" at the annual garden school of the Raleigh Garden Club, before which his friend and colleague Z. P. Metcalf also appeared. In February 1937 he and Metcalf led a roundtable discussion at a conservation conference sponsored by the Garden Club of North Carolina, and the following month he served as instructor at its state flower show school in Chapel Hill. He also spoke at the fourth conservation conference held by the club in 1941. From 1937 until at least 1942 he was listed as a participant in the speakers bureau of the garden club. The titles he offered were at first "Beginner's Botany" and "Plants of the Great Savannah"; in 1942, following his honeymoon trip, he also listed "An Ecological Cross Section of Mexico." His apparently final appearance for the organization was before the Cape Fear Garden Club in 1951. On this occasion he spoke about the geological history, ecological situation, and vegetation of the Wilmington area; he particularly emphasized Greenfield Lake, Bald Head, the Big Savannah, and Fort Fisher. By this time, he had the advantage of color photography, and a listener reported that "both the unusually clear slides and Dr. Wells's fascinating side-lights brought out the varied selection of plant life to be found in all these areas."[4]

Ever interested in trying new things, Wells extended his popular talks to the then new medium of radio. During the years 1935 to 1944, the North Carolina Agricultural Extension Service produced a fifteen-minute daily program. Entitled "Carolina Farm Features" and aired by station WPTF and later station WRAL, this program originated live from the State College campus and featured a six- to

B. W. Wells, at center, on one of his many field trips with students and laypersons, about 1926. Courtesy of Maude Barnes Wells.

eight-minute talk prepared and delivered by some specialist at the college. One department at a time was assigned responsibility for the talk for a given day each week. For more than a year during this period, the Department of Botany had that weekly responsibility; various members of the faculty, including D. B. Anderson, I. V. D.

Shunk, and L. A. Whitford, participated in the program by discussing subjects related to their scientific specialties. Already practiced in such popularization, Wells spoke about ten times on ecological topics and on wildflowers.[5]

Wells continued his speaking activities until late in his life. Most of his talks in the later years of his professional retirement were delivered to the North Carolina Wild Flower Preservation Society, usually on occasions of the semiannual meetings of that group. In October 1955 at Umstead State Park his talk was titled "The Distribution of Wild Flowers in Relation to Plant Succession." In May 1963 he entertained the society at his farm home and discussed the flora of that tract. In October 1964 in Fayetteville he delivered "The Future of Wild Flower Preservation," and in May 1966 he gave a brief account of the history and vegetation of the sandhills region. Finally, in May 1967, at the age of eighty-three, his talk at Old Brunswick Town was called "Some Plant Communities and Habitats of the Remarkable Wilmington Area"; the vegetation he discussed included the seaside dune, shrub, and salt-spray communities, the shrub-bog pocosin and Carolina Bays, the sandhill longleaf pine forest, and the vanished Big Savannah.[6]

Wells's lectures were well received and in demand over many years because his ideas were stimulating and his delivery forceful. Moreover, he had the rare ability to establish rapport with nonscientists and communicate technicalities in everyday language; as one observer put it, "Though his talks are quite scientific, he always gives them in terms that we can understand and in such a manner that his enthusiasm is contagious." The number of public lectures Wells delivered during his career cannot even be estimated, but certainly it was large. In one fifteen-year period, from 1920 to about 1935, he judged that he gave more than one hundred such talks to garden clubs, home demonstration clubs, civic clubs, various societies, and public schools. He traveled extensively in this endeavor but always charged only for expenses, if at all. The full impact over the years of his popular talks cannot be assessed, but almost certainly it was enormous. Through such lectures he became known to and respected by a great many citizens of North Carolina. More important, through such lectures a great many of those citizens were stimulated to appreciate more fully the natural beauty of their state.[7]

Wells could stimulate and inspire with a lecture, especially an illustrated one, but he particularly liked to show his listeners real plants as he talked; such exposition, of course, usually required a trip to the field. Yet the first occasion on which he participated in demonstrating plants of North Carolina to its people involved bringing the outdoors in, rather than taking those people out. In the spring of 1920, less than a year after he first arrived in North Carolina, Wells collaborated with Susan Franks Iden, a reporter for the *Raleigh Times*, in staging a wildflower show. The two brought in freshly gathered spring flowers and arranged them in a public display on the second floor of the Olivia Raney Library in downtown Raleigh. A contemporary photograph shows the flowers artistically grouped in individual baskets, which were arranged around the room on tables and the floor. This exhibit caught the eye and fancy of Charlotte Hilton Green, a writer on subjects of nature and conservation, who, like Wells, was newly arrived in Raleigh. Green was later to become well-known throughout the state for the newspaper column she wrote for forty-two years, "Out-of-Doors in Carolina." She composed an article about the flower exhibit, which she titled "When the Wild Flowers Came to Town"; the national magazine *Ladies' Home Journal* purchased her piece for the sum of eighteen dollars but never published it.[8]

Susan Franks Iden, who collaborated with Wells in staging this wildflower show, was something of a pioneer as a female journalist in a southern town. A Raleigh native, she had a bent for writing, which she eventually turned into a career. She was very active in the Edenton Street Methodist Church and in 1912 wrote a history of its Sunday school to mark the opening of a new building. In 1920 she published a detailed description of the tercentenary pageant staged by the people of the town to commemorate Sir Walter Raleigh and his dream of a New World colony. This large-scale production was a community effort performed in the local baseball park on three nights during the week of the state fair. Iden worked as a reporter for the *Raleigh Times* from at least 1920 until at least 1933; she also occasionally furnished articles to the *Charlotte Observer*. Her interest in plants and in writing about them drew her to Wells in a friendship that lasted for several years. Their association bore its greatest fruit in the circumstances leading to the publication of Wells's book, *The Natural Gardens of North Carolina*. In addition to playing a crucial role in the sponsorship of that book by the Garden Club of North

Carolina, Iden also furnished one of the photographs in it, that of the flowers of the bloodroot plant.[9]

In telling others about the plants he loved, Wells was at his best when confronting nature outdoors, speaking extemporaneously in the midst of the actual vegetation and habitats. On a field trip he was superb; until he reached an advanced age, he was eager to lead a party out and, having arrived, to talk at length in a way his followers found spellbinding. The number of such trips he conducted during his long career cannot be reckoned, but it must have been large. In addition to the excursions associated with his formal classes for college students, he led numerous trips for laity and professional biologists alike. In 1935, for example, he led the Clinton Garden Club on an expedition through White Woods, a site near that town which included most of the vegetation types characteristic of the coastal plain; he ended the excursion with a lecture called "The Soil Control of Plant Distribution." The itineraries of Wells's trips usually consisted mainly, but not exclusively, of sites at which he had conducted research investigations, and so they varied over the years. But according to available accounts, they apparently differed little in tone and spirit. They usually involved hurried travel between points of interest, strenuous physical activity and intellectual challenge at the sites studied, and some well-orchestrated social interlude.[10]

A detailed description exists of one of Wells's early field trips, taken in 1926, which was probably typical. Written by Susan Iden, the newspaper account was topped by the full-page heading "North Carolina Makes Another Claim to Greatness"; in the style of the day, subheadings declared, "Nature Study Trip Reveals Wonders of Coastal Plain Region," and "Dr. B. W. Wells of State College Conducts Party from Raleigh to Carolina Beach and Fort Fisher with Big Savannah and Its Pageant of Wild Flowers Climax of Interest." The article included three illustrations: a view of the wildflower expanse of the Big Savannah, a close-up picture of white fringed orchids, and a shot of the party searching for Venus flytraps. On this trip a group of fifteen persons left Raleigh in a caravan of five Model-T Ford automobiles for a two-day tour of Wells's favorite haunts in the coastal plain. The party included college students from both a class in botany and one in science, as well as two laypersons, one of whom was Iden. Wells emphasized the topography of the landscape as well as the vegetation; as the trip progressed, he made several stops to point out the series of terraces that make up the coastal plain. Plant communities visited on the first day were a beech and maple forest along Washington Creek, a pocosin (shrub-bog or bay), and the Big Savannah. The party punctuated the trip by an overnight stay, some in a shack on the savannah and some at the hotel in Burgaw. On the second day, the itinerary included visits to swamp forest and aquatic communities along the Cape Fear River, a freshwater marsh near Wilmington, a sand-ridge turkey oak stand, a yaupon community at Fort Fisher, and a salt marsh near Carolina Beach. The tour included lunch at Wilmington with a visit to the Coast Guard cutter *Modoc*; it ended, characteristically of Wells, with a beach party around a fire of driftwood.[11]

Another account, written by Wells himself, describes a trip which he conducted in April 1938 for a mixed group of scientists and others. The twenty-four participants, fourteen males and ten females, represented the Southern Appalachian Botanical Club and the Torrey Botanical Club; they included two persons from Washington, D.C., four from West Virginia, five from New York City, six from North Carolina, and seven from Ohio. Probably because of the interests of the party, Wells on this tour paid more than usual attention to rare or striking occurrences of individual plant species: an orchid at Lake Waccamaw, rosemary at Myrtle Beach, and the sandhills selaginella at Carolina Beach. But in the main he led the party to the usual plant communities: a wiregrass savannah, the shrub-bog pocosin of Angola Bay, a dune-shrub community, a gum swamp forest, and the Big Savannah. The tour also included visits to Brookgreen Garden in South Carolina and Greenfield Park near Wilmington, as well as an evening social reception at the home of Janet Bluethenthal.[12]

Wells also provided a written account of a foray which he led in 1953 for a thoroughly professional party from the Southern Appalachian Botanical Club and the Association of Southeastern Biologists. This group of twenty-four males and nine females included twelve persons from North Carolina, seven from Georgia, six from Virginia, three from Tennessee, two each from Alabama and Pennsylvania, and one from Florida. The company traveled in a convoy of fifteen cars led by a state highway patrolman. The latter, Wells wrote, "saved the party much time by the elimination of traffic light stops"; he strongly recommended such an arrangement whenever possible. The sites and plant communities visited included those which by now were standard for Wells: the turkey oak and wiregrass community of the sandhills, White and Singletary lakes, the dune community at Fort Fisher, the shrub bog at Holly Shelter Bay, and, of course, the Big Savannah.[13]

Wells led many more field trips than those just described; each year, for example, he scheduled at least one long tour for his college class in ecology. But whoever the clientele, a Wells-conducted trip was a memorable event. Most persons who participated in one never forgot the experience, which usually expanded their views of nature by focusing their eyes on hitherto unrecognized details. Looking back twenty-five years, the botanist Herbert Temple Scofield said, "They were wonderful because of his great enthusiasm and energy. He had an outstanding knowledge of eastern North Carolina and a tremendous rapport with the students." Richard A. Popham, a professional botanist from outside the state, who participated in only one outing with Wells, could recall with great clarity and in extensive detail the sights and occurrences on a trip taken more than a quarter-century earlier. And a nonscientist wrote in effusive terms: "Throughout these many decades what a guide and leader he has been to so many—students, classes, Garden Club, Bird Club members, and what-have-you. . . . So many, many trips and how interesting he made them all!" Wells was indeed a voice for and demonstrator of North Carolina nature.[14]

CHAPTER 6

Popular Pen

To us tiny creatures of intellect, who, bacteria-like, inhabit the threads of this great green quilt, the patches are very clear to us if we are observant on our rambles. On the train or by auto, from Raleigh to Wilmington, for example, one may get close-ups of some of the most interesting and diverse vegetational masses in all North America. Or if one begins at the summit of Mt. Mitchell and by auto winds his way down to its base and then eastward to the Cape Fear peninsula, taking passage there to Smith's island, he will, if his eyes are open, see one of the most remarkable vegetational panoramas in the world. Not many geographical areas anywhere can equal such amazing contrasts within its borders. It behooves every North Carolinian to know North Carolina first.—Bertram Whittier Wells, 1924

IMPORTANT AS Wells's lectures and field trips were in popularizing plants and vegetation, the writings he directed at the public were even more so because the audience they reached was much larger. The supreme product of such activity, of course, was his book, *The Natural Gardens of North Carolina*, but that striking achievement tops an impressive list of lesser works. Before he came to North Carolina, he did not produce such pieces, concentrating on his scientific and academic activities. But after settling in the state, and especially after he had surveyed its vegetation, he contributed steadily to the popular and semipopular press. During the first fifteen years of his residence in North Carolina, he produced several newspaper articles; unfortunately, no detailed list of them exists. In the summer of 1925, however, he published in the *Raleigh Times* a description of the Big Savannah, which he dubbed "God's Garden of Eastern North Carolina." In about 1935 he listed the subjects of some newspaper accounts he had authored; these included homelands of garden flowers, shrub bogs of eastern North Carolina, the Venus flytrap, old-field weeds, plant succession, autumn tints, mountain balds, grass and human society, the corn plant, and the cotton plant.[1]

Wells's bent for popular writing harmonized with the plans of Eugene C. Brooks, president of North Carolina State College from 1923 to 1934. When Brooks assumed office, he began a campaign to publicize the institution and its work in technical fields. The college Extension Division was expanded; Wells's service as a speaker for this agency has already been cited. Created at this time as part of the public relations effort was *North Carolina Agriculture and Industry*, a one-page weekly information sheet that tooted the horn of the college. Printed in large-page newspaper format, this publication was widely distributed through three volumes of twenty-seven numbers from 17 October 1923 to 27 May 1926. Wells contributed to it.[2]

In the first issue he wrote about the economic possibilities of a beverage made from leaves of the yaupon or cassina plant. This shrub had long been used for such a purpose by Indians and others and was effective because of its flavor and caffeine content. Basing his article on a recent bulletin of the United States Department of Agriculture, Wells pointed out that the possibilities included use of the plant in preparing both a hot tea and a cold drink, bottled or from a drugstore fountain. Because the plant grew extensively in sterile sandy soils along the coast, exploiting it in this way would create a totally new economic resource for the state. His optimism fairly bubbled: "If the public can be educated to the use of the new drink (a matter of advertising purely) there can be little doubt that North Carolina will have the pleasant surprise of finding an additional and little-dreamed-of source of income, all of which is just another sample of the underdeveloped resources in this vastly variegated state, a commonwealth simply replete with potentialities."[3]

In another article, which appeared a few weeks later, Wells described a summer camp held in 1922 by the School of Forestry of Iowa State College. He encountered this operation on a field trip while looking for a camp site in the Pisgah National Forest. Naturally he inquired as to why seventeen students and two professors from Iowa were pursuing their work in North Carolina; the answer of a professor was that "I can teach more and better forestry than anywhere else." Wells drew a moral that must have gladdened President Brooks: Iowa, which had few forests, sent its students to study in a state that had no forestry school. "Now these facts should jar every North Carolinian," Wells wrote; "the youths of North Carolina who would like to begin a forestry career and serve the great North Carolina forestry interests now have no opportunity afforded them by the State." Although courses in the subject had been taught at State College since 1892, at the time of Wells's article President Brooks was pushing hard for establishment of a school of forestry. His efforts began to pay off in 1928, when a four-year degree curriculum was approved, and in 1929, when a Department of Forestry was created in the School of Agriculture. A separate school did not appear until long after Brooks's time, but it did come.[4]

Wells's principal contribution to *North Carolina Agriculture and Industry* was the written text of his lecture "The Patch Work of North Carolina's Great Green Quilt." This work was serialized in six installments bearing seven photographic illustrations. It described the eleven major plant communities he had recognized in his recently completed technical bulletin: the sea oats of the dunes, the rush and grass salt marsh, the cattail and sedge freshwater marsh, the pondweed and waterlily aquatic community, the orange grass and trumpets savannah, the gallberry bay land, the gum and cypress swamp forest, the turkey oak and wiregrass sand ridge community, the oak, maple, and pine upland forest, and the fir and spruce mountain forest. But the descriptions were framed in the figure that the vegetation

B. W. Wells with his first wife, Edna, and their young protégé Frank Johnson, relaxing on a field trip near Carolina Beach, 24 June 1922. Courtesy of Maude Barnes Wells.

of the state was a "great green quilt" on which the plant communities were recognizable "patches."[5]

In approach and style this work clearly foreshadowed *The Natural Gardens of North Carolina*, which appeared eight years later. The language was typical of Wells at his most enthusiastic. He noted that the Big Savannah was "as level as a dance floor," and he characterized the spruce and fir forests as "vegetational splotches which the long cold finger of Boreas has daubed on the summits of our high mountains." His love of the natural outdoors tumbled forth in statements like "To spend a day in a dugout, drifting idly amid the silent shades of a great cypress or gum swamp is an experience never to be forgotten." Similarly, he confessed that in the primeval forest of the mountains "there breathes out continually that subtle, deep woodland aroma which catches at the nature-loving instincts of every human being."[6]

But Wells did not obscure the principal aim of the work, which he clearly stated: "It behooves every North Carolinian to know North Carolina first. . . . Every intelligent North Carolinian should be able to back up his pride by facts and figures—these always count most." That he himself in mind and spirit was already totally one with outdoor North Carolina is revealed by his conclusion:

> Now in the way of summary let me say that it is a thrilling experience to stand in the forest warden's tower on Mitchell amid, yet above, the deep green balsams and spruces, and facing east look beyond and down on the great broad-leaved forests on distant slopes, thence in imagination go beyond the blue ranges to the sunny fields and ranges of the Piedmont, on to that marvelous medley of the Coastal Plain—the rich verdure of river slopes, the singing pines, the suffering oaks of the burning sands, the dark swamps, gray with "moss," the brilliant, multicolored savannahs, the impenetrable pocosin thickets, the quiet waters of pond weeds and water lilies, the vast fresh and saltwater marshes, and at last the dunes, with their grass waving under the eternal wind from the sea. Nowhere in all Eastern America has Nature bestowed such a wonderful heirloom as this marvelously variegated great green quilt which is spread over North Carolina.[7]

This piece of work evoked favorable public comment, especially by comparison with the usual offerings of publicity sheets like the one in which it appeared. "We have welcomed the sight of and thoroughly enjoyed" the article, one local newspaper declared; "even the suspicion that one is being instructed does not detract from one's pleasure in coming across in the great mass propagandic collegiate printed matter something that neither affects to save the state nor brags on its high birth rate." These comments were reprinted and seconded by the student newspaper at North Carolina State College.[8]

President Brooks of State College, in addition to directing his institution, was active in other state affairs. He served as secretary of the North Carolina State Park Commission, established in 1924. In that capacity he played a key role in the events that led to the creation of the Great Smoky Mountains National Park. This park, established in principle in 1926 and finally dedicated in 1940, was one of the major early conservation projects in the eastern United States. It was a venture Wells warmly supported, so when Brooks asked him to prepare a botanical brochure to be used in influencing the North Carolina legislature, he gladly complied. At some time before 1931 he produced the booklet *The Remarkable Flora of the Great Smoky*

Mountains. Later Susan Iden used this work in her efforts to interest publishers in Wells's projected book.[9]

Wells also contributed sections or chapters to books edited by others. In 1927 an article by him on southern wildflowers appeared in an encyclopedia for children. And a year earlier he and Z. P. Metcalf had collaborated on a description of North Carolina for the *Naturalist's Guide to the Americas*. This volume was intended to itemize areas that remained in their natural condition. In an introductory discussion, Metcalf provided descriptions of the characteristic animals of the state, and Wells summarized the plant communities to be found. They also listed the principal sites of ecological interest which were under some kind of protection, with specific directions for location. Many years later, in 1949, Wells prepared a section on North Carolina wildflowers for *A Traveler's Guide to Roadside Wild Flowers, Shrubs, and Trees of the U.S.*, a publication sponsored by the Garden Club of America and the National Council of State Garden Clubs. In this piece he briefly listed characteristic native plants of the state, examples of agricultural land use, and centers of floral interest. In 1950 he also contributed a section on flora and fauna to John A. Oates's *Story of Fayetteville and the Upper Cape Fear*; in this discussion he emphasized the contrast between the original longleaf pine and hardwood forests and "the complex series of small communities which have followed the lumberman's saw, the farmer's plow, and everybody's fire." A second edition of this book was published in 1972 and a third in 1981.[10]

One such contribution by Wells to a larger book deserves closer attention. In 1939 he wrote an unattributed section called "Natural Setting" for *North Carolina: A Guide to the Old North State*. This book was one of the state guides undertaken as a relief measure during the economic depression of the 1930s; its preparation was undertaken by the North Carolina branch of the Federal Writers' Project of the Federal Works Agency of the Works Progress Administration, the WPA of the New Deal governmental alphabet soup. The book really was, as the title claimed, a comprehensive guide to the state. It provided extensive general background information, described the major cities, and presented detailed directions for highway tours. Two of the tours included touches of Wells: in one the Big Savannah was noted as a point of interest, and in another his hypothesis of Indian origin of the mountain grassy balds was mentioned. In his own chapter Wells discussed the physical geography, climate, plants, animals, and natural resources of the state; the latter included forests, minerals, water power, and fisheries. The section on flora occupied about 15 percent of the total. In 1955 the book was revised as *The North Carolina Guide*; and though the work as a whole was enlarged by forty-eight pages, about 8 percent, Wells's part was expanded by seven pages, about 30 percent. Wells reorganized his chapter and added new sections on geology and on lakes of the state, the latter affording him an opportunity to indulge his current interest in the ecology and history of the Carolina Bays. But most of the expansion occurred in his discussion of plants and vegetation, which was increased more than threefold; the viewpoint was also changed somewhat. The revised discussion began with a lengthy quotation from *The Natural Gardens of North Carolina* and emphasized more detailed ecological interpretation. "The best way to interpret the flora of a State," he wrote, "is to deal with it on a community basis, as the more prominent and stabilized community types are correlated with diverse habitats." He described seventeen community types, as follows: rock communities of lichens and mosses, old-field communities of herbaceous weeds, dry woodlands (mainly pine), deciduous forests, aquatic communities, freshwater marshes, wet woodlands of river floodplains, swamp forests, shrub bogs, savannahs, salt marshes, sandhill communities, seaside or salt-spray communities, balsam and spruce mountain forests, mountain fire cherry and red elder, high-mountain shrub balds, and high-mountain grassy balds. Through all his description there flowed his recurrent theme of the wide diversity possessed by North Carolina in climates, soils, and vegetation.[11]

From time to time Wells also wrote on science for the North Carolina magazine the *State*. After a storm devastated much of Carolina Beach in August 1944, he produced an article aimed at correcting the public impression created by news accounts that the damage was directly attributable to high wind speed. He discussed the factors involved and pointed out that losses from the wind itself were minor, the major culprit being strong shore currents elevated by the high winds. "What really happened," he wrote, "was that the shore current acted as a giant carving tool cutting the upper beach back 10 to 20 feet and 1 to 3 feet deep. . . . Beach front property owners found they had 10 feet clipped off their lots and a foot added vertically, which, unfortunately, is not good real estate practice." He also explained that the injury to trees and other plants observed within a week up to ten miles from the ocean was the result of salt spray carried from the waves by the high wind. Later he also contributed to this magazine a lengthy letter about the mountain grassy balds in which he discussed in popular terms some recent scientific work by others and outlined his own hypothesis of the Indian origin of these vegetational features.[12]

One article Wells wrote for the *State* is interesting because it reveals much about some of his personal characteristics and his complete and consistent dedication to a scientific view of the whole world, not just of the plants he studied professionally. This piece was an engrossing description of his attempts, at first methodical and ultimately somewhat frantic, to explain some "ghostly" sounds heard in his house at Southport. This building, located on the bank of the Cape Fear River, was a handsome and very old structure known as the Stuart House. Soon after Wells and his wife, Maude, had occupied this summer house for the first time, they began to be awakened in the night by strange, muffled, rumbling sounds. These sounds were only seconds long and might be separated by hours of silence; they were never heard during the day. His account describes his mental state in approaching such a problem as a scientist and his emotional state as his efforts to solve the mystery continued to fail. Those efforts included sleeping in different rooms in an attempt to fix the source of the sounds, investigating noises made nearby by men dragging shipping crates over a rough floor, and marking the locations of trunks in the attic to detect their possible movement. After two months of puzzlement, his distress at a feeling of helplessness in the face of the unknown was acute: "My whole mental set-up from boyhood was that of belief in and trust in the natural world and now the natural world in this particular spot was deserting me. I can never forget the unhappy even fearful state of mind it produced. I found it difficult to concentrate on my field scientific work and when on an occasional sleepless night the old rumbling came back again, I felt like a man with his back to the wall facing an unknown but real enemy (ignorance) and now I was incapable of doing anything about it." In fact, the explanation was simple. He ultimately discovered it by chance rather than scientific design, but the acute awareness of the practiced observer sensitive to his surroundings was involved. "By the sheer accident of timing and proximity," he wrote, "and not through any logic of science, this mystery collapsed." The culprit was a piece of sheet iron used to close a fireplace opening: this metal acted as a sounding board for the fluttering of the wings of a colony of swifts which had nested in the chimney, unknown to Wells. His relief was immense, and his faith was restored: "For me a great dark cloud rolled from my mind and I regained my trust in the natural world."[13]

Wells wrote for readers of varied backgrounds but almost always with the aim of relating the vegetation and plants of North Carolina to their interests. In a brief article for a publication devoted to birds, for example, he discussed some relations of those creatures to the vegetation around Southport. And in his later years he prepared some short articles for the newsletter published by the North Carolina Wild Flower Preservation Society. One of these correlated the occurrence of wildflowers with community stages in plant succession, another described the extensive flora of the 150-acre farm to which he had retired, and still another was the text of a talk he delivered to the society at Old Brunswick Town. His last article, produced in 1971, when he was eighty-seven years old, was a discussion of the transplanting of wildflowers from their natural sites to home gardens and yards; he felt this to be an extremely important activity because "millions of children are growing up with no knowledge of the wild flowers, which were so familiar to their ancestors."[14]

One technical idiosyncrasy of Wells's should be pointed out. In his popular and semipopular writings as well as in his talks, he invented common or vernacular names for plants; this fact is supported by the testimony of two of his longtime colleagues. Plant names, of course, are arbitrary, as Shakespeare's Juliet pointed out so lyrically for the rose. But botanists have had to be systematic about names and so have developed a system of botanical designations formulated according to precise rules and uniformly understandable to themselves. Every kind or species of plant has such a name; unfortunately, that name has the form and ring of medieval Latin, nowadays a foreign language to almost everyone. Of course, many botanical names have been absorbed into English and are familiar and easy to use; some examples are asparagus, chrysanthemum, geranium, and petunia. And over the ages many widely familiar plants have acquired common or folk names in the language of the people; examples are oak, maple, toadflax, chickweed, and Johnny-jump-up. Many plants, however, especially wild ones, have acquired no popular names that are in general use. Thus the person interested in plants may be confronted by such technical botanical labels as *Spermolepis divaricatus* or *Podostigma pedicellata* or *Tetragonotheca helianthoides* and may consequently become discouraged. Wells aimed to make botany as simple as possible for the nonprofessional and believed the plants themselves were the important things, not their names; he thought that those names should, if possible, be aids to the memory. And so if a plant worthy of notice did not have a widely accepted everyday name, Wells simply created one.[15]

Examples of this practice are scattered throughout Wells's book, *The Natural Gardens of North Carolina*. In many cases he used common names that were different from those employed in the leading technical botanical manuals of the time, especially those by the expert on the Southeast, J. K. Small. Three examples are "scarlet-flowered bean," called by Small "cardinal spear" (*Erythrina herbacea*); "red hot poker," to Small "wild bachelor's button" (*Polygala lutea*); and "loud speakers," Small's "Barbara's-buttons" (*Marshallia*). It is difficult to say whether Wells used these different names because he felt they would be easier or because he thought they better reflected local usage in North Carolina. In other cases he employed common names for plants that received no such designations at all in the technical manuals; "pea tree" (*Sesbania*), "thread plant" (*Burmania*), and "hide-and-seek" (*Zornia bracteata*) are examples. In two such cases it is certain that the names were Wells's inventions. He once received an inquiry from a woman in California who wished to acquire some *Centella*, a small plant to be found near the coast; she had lived in India, where the plant was eaten for a healthy life and high intelligence. Wells collected and sent her some and subsequently used for it the name "intelligence plant." In another case, he definitely coined the name "green-and-gold" for *Chrysogonum virginianum*, an attractive wildflower. Unfortunately or not, depending on one's viewpoint, the latter name is the only one floated unilaterally by Wells which has been used in the currently authoritative technical manual for the region.[16]

A dedicated expository writer on technical subjects for the public aims for his products to be read, understood, and if possible enjoyed. Wells's were. He was not remarkably prolific as a writer of popular pieces, but his interest was continuous and the stream was steady if not voluminous. His life in North Carolina spanned a little over fifty-nine years, from age thirty-five to age ninety-four; during that time his popular writings appeared over a period of fifty-one years, from age thirty-six to age eighty-seven. What is notable is not that he did not produce more, but rather that in the face of his other activities and circumstances he produced as many as he did. Even more important is the fact that a significant readership found his efforts both instructive and interesting.

CHAPTER 7

Natural Gardens

As I am still unable to write since my illness this fall my friend is writing this for me, to tell you how delighted I am with the book. Dr. Wells came to see me Sunday afternoon and presented me with a complimentary copy which has scarcely been out of my hand since that time. In addition I have two copies which I bought to use as Christmas gifts. The book far surpasses even my fondest dreams and I think the University Press has done an excellent piece of work. I hope you are as greatly pleased with the production as I am and as Dr. Wells seems to be. I hope it will work out a financial success for the University Press. Certainly it is a great contribution to the educational life of the state.

—Susan Franks Iden, 1932

WITHOUT QUESTION, Wells's greatest and most enduring communication with the public occurred through his book, *The Natural Gardens of North Carolina*. This work was a highly successful attempt to popularize ecological concepts and their relation to wildflower occurrence and conservation. First published in 1932, it has over the years brought both information and pleasure to a wide audience. Today it is still useful and still used.[1]

The Natural Gardens was unique among wildflower books in its ecological orientation. Wells organized his material into two parts, "The Natural Gardens of North Carolina" and "The Herbaceous Wild Flowers of the Natural Gardens." The first consists of descriptions of the eleven plant communities he had recognized earlier in his scientific research; the second contains simplified keys for identifying the principal wildflower plants, together with descriptions of those plants. He couched his discussions of the communities in nontechnical but accurate terms. As he had done before, he invented popular names for them; examples are "deserts in the rain" for the sandhills community, "the melting pot" for old-field communities, and "Christmas tree land" for the forest of the high mountains. The text was extensively illustrated by photographs and diagrams.

The book is a tour de force. Its language reflects Wells's personal spirit, teeming with effusive description: "The scene of golden goblets scattered over a vast green festal board would not compare with that of the trumpet flowers dotted over the lawn-like expanse of the early spring savannah." Some of its figures are extreme: "Like the ruthless gangster he defies society successfully. He is seldom caught, for his allies are too strong. His name is Fire." Some of its allusions are dated: "Like political parties, our bogs differ in degree of wetness." Wry observations abound: "Unlike mosquitoes, weeds are not wholly bad"; "Variable spurge (*Tithymalopsis Ipecacuanhae*). Perhaps a better name would be 'the long name spurge,' for its technical appellation is always of interest to budding biologists, many of whom will learn it when shorter names will be totally unremembered." But the book is salted with homely metaphor and peppered with down-to-earth practicality: "Hence it is that bogs are, in a sense, one of nature's canning factories, where plant products are preserved"; "In the great sandhills every illiterate countryman knows that black jack oak (*Quercus Marylandica*) will tell you where the clay layers occur at or near the hill tops"; "If white oak and dogwood are there, let go of the money; the land will be the best the sandhills can give one." And it is replete with simple-hearted appreciation of nature: "To know exactly in the woods where Solomon keeps his seal and where the Indian finds his pipe is very worth while knowledge—if one is still youthful enough to appreciate such valueless information"; "The name may come last or even not at all; the knowledge of and aesthetic joy in the plant itself is the all important matter."[2]

It might be thought that the reader of today would find such expression too cloying; it smacks, in fact, more of the year 1900 than 1932. But Wells never patronized, and his evident sincerity and enthusiasm for his subject save the situation. He wrote with the simple exuberance of an adolescent and the acute wisdom of a savant, and the reader is swept along to enlightenment.

The book met a good reception from the press, both popular and scientific. The *Biblical Recorder* noted that "it is not a technical book, but is written in such a way that a person who has never studied botany can understand and appreciate the beauty and great variety of our native plants." Jonathan Daniels, writing in the *News and Observer* of Raleigh, called it an excellent book "altogether both in its literacy and its scientific character" and recommended it strongly to "every North Carolinian who loves to dig in the earth or who takes joy in the loveliness which grows out of the earth of his State." He attributed to Wells "the rarest ability in translating scientific enthusiasm into a graceful and popular book." The reviewer for *Science News Letter* stated, "Seldom has the vegetation of a state received a treatment at once so adequate botanically and so interesting to the nonprofessional outdoorsman." *American Forests* asserted that "it amply fulfills its declared purpose" and called the book "a fascinating one . . . filled with striking habitat photographic illustrations." Writing in the scientific journals *Ecology* and the *Botanical Gazette*, the plant ecologist G. D. Fuller applauded Wells for fulfilling the generally neglected obligation of the scientist to offer the layman well-selected facts accurately stated. Fuller objected mildly to the teleology of many of the descriptions but concluded that "the personification of trees and flowers, while objectionable to the scientist, may be pardoned when it accomplishes its object in enlisting the interest of the non-scientific reader." He thought that the book would carry the results of specialized ecological studies into the homes of many laypersons and would provide a model for other universities to follow.[3]

B. W. Wells near Carolina Beach, using the unwieldy camera with which he made hundreds of photographs to illustrate his many popular lectures and his book, *The Natural Gardens of North Carolina*, 1926. Courtesy of Larry A. Whitford.

Wells aimed his book well and hit its target simply and directly. But the story of how *The Natural Gardens* came to be written is anything but simple and direct. It had a dual genesis. First and foremost, of course, was the development of Wells's own knowledge of the plants and vegetation of the state, along with his enthusiasm for telling others about them. His lecture and article "The Patch Work of North Carolina's Great Green Quilt" was the clear forerunner. But the book probably would not have been created without the inspiration and practical efforts of two members of the Garden Club of North Carolina, Ethel D. Tomlinson and Wells's friend Susan F. Iden. These two envisioned such a book and stimulated Wells to bring it to fruition.

Tomlinson, wife of a well-to-do furniture manufacturer in High Point, became president of the garden club in 1929. She had already recognized the need for a book on native North Carolina wildflowers because of numerous requests for information received from local clubs and schools; no suitable source was available in print. When she became president, she made it a special aim of her two-year administration to bring about the publication of a bulletin or book on the subject. During the second year of her term, 1930–31, she created a conservation committee chaired by herself, with such a publication as its sole objective. For assistance she turned to Iden, who became conservation chair in 1931, because the latter over a period of years had written various newspaper articles about nature and wildflowers and because her column "Trailing the Wild Flowers" was then appearing in the *Charlotte Observer*.[4]

Iden wasted little time. She quickly ascertained that indeed no suitable source was already available and that none would be forthcoming from the state Department of Conservation and Development. Because of her previous contacts with Wells, who had imbued her with enthusiasm for the ecological relations of plants in their native habitats, she turned to him; he agreed wholeheartedly that the need existed and expressed a willingness to undertake the task of writing a suitable book. By January 1931 she could report to the executive board of the garden club that Wells had agreed to do the work and would accept no compensation whatever for it. He apparently also wasted little time: he had already sketched a plan of the work and had composed two chapters, parts of which Iden read to the board. After discussing how the book might be financed, the board voted to endorse the project and to recommend its undertaking at the annual meeting of the club scheduled for April.[5]

Iden then began to search for a publisher. She first communicated with the Macmillan Company of New York. That firm was interested, but under the terms of its proposal the book would have to sell at $5 per copy, a price she thought too high. Next she discussed the project with William Terry Couch, assistant director and later director of the University of North Carolina Press in Chapel Hill. This establishment, organized in 1922, was still in the fledgling stage. It had already published 120 books and several journals, but because of economic conditions at this time was experiencing severe financial constraints. Nevertheless, Couch responded favorably, but with one crucial condition: the press would publish the book in an edition of two thousand copies, provided the garden club would subscribe to five hundred of the copies at a price of $3 each. The press expected to sell the remaining copies mostly to schools but also to an estimated two to three hundred individuals.[6]

On 22 April 1931 Iden reported her findings to the executive board of the garden club, which discussed them without taking definite action. The next day Tomlinson and Iden laid the matter before the club as a whole at its annual meeting in Wilmington. They reported that Wells had offered to write the book without compensation if the club would help finance it, that he would give up a planned trip to the West if the club would secure publication before Christmas, and that he estimated the book would contain about three hundred pages and about 150 pictures, some in color. After hearing the proposals made by both Macmillan and the university press, the thirty-five members attending voted unanimously to accept Wells's offer and undertake publication on the university press's terms, with the specific means of financing the project left to the executive board. Tomlinson had made every effort to ensure the club's approval of the project, for she had arranged for the dynamic Wells to speak to the club that night before the business session; he repeated this presentation for some local clubs the following June in Greensboro.[7]

The stage was thus set for Wells to write the book and for the garden club to arrange its share of the financing. The accomplishment of these two goals now became two threads interwoven into the fabric of one story. Initially there was optimism on all sides that publication could occur by Christmas 1931. In the April report of the conservation committee Iden not only described the approach the book would take and presented an outline of it, she also expressed the possibility that the work might go to the publisher by the summer. Tomlinson assumed the office of publications chair; she took charge of the efforts to raise the necessary $1,500 through advance subscriptions to garden club members and was sure enough of the outcome to tell Wells to proceed. A payment schedule was agreed upon: the club would remit to the press three installments of $500 each, the first when Wells submitted the manuscript, the second when he returned corrected galley proofs, and the last within thirty days of actual publication. By June the garden club had collected $250 in advance orders for the book, and more came in throughout the summer.[8]

For his part, Wells was also moving ahead; he intended to complete the book during the summer. As early as June, however, he began to realize that he could not get the manuscript into final form in time for Christmas publication because it would take some months to polish it and make ready the necessary illustrations. Nevertheless, he worked hard during his entire vacation from classes. But the deadline could not be met, and publication was postponed. During the school year of 1931–32, Wells continued to work on the book on Sundays and holidays, and during the spring he tested the flower-identification keys by putting them to use in his botany classes at the college and in the high school biology classes his wife taught. Iden also contributed at this time by reading parts of the manuscript and by trying out the keys. By June 1932 the rough draft was almost complete and the photographs were ready.[9]

But a new obstacle now arose. The stock market crash of 1929 had brought on the Great Depression, which by 1932 had reached disastrous proportions. Banks failed throughout the land, and one of them, the Commercial National Bank of High Point, held the $587 which the garden club had so far collected for the book. By March 1932 these funds were frozen indefinitely and assumed to be totally lost. Tomlinson, however, would not give up. At its annual meeting in April the garden club voted to give *The Natural Gardens* priority over all other projects and instructed Tomlinson to press on with the financing. She started a new account in a different bank and managed to collect enough money that the available funds grew from only $84 in April to $750 in July. By the latter time she believed that the entire $1,500 could be raised by the end of the year, and the university press was willing to trust that this was so. Therefore, the two parties agreed that the project could proceed as soon as Wells had readied the manuscript.[10]

At this point, however, a misunderstanding arose which might have derailed the whole undertaking. The club had originally expected publication to occur by the previous Christmas; when Wells had found that he could not adequately prepare the manuscript in time, the club had asked its subscribers to wait and had solicited orders for additional copies with the promise that the book would become available by the end of 1932. Tomlinson felt that delaying longer might destroy the club members' interest and wreck the project; therefore, she thought the matter should be pushed. Wells, however, from his conversations with Iden, was under the impression in June that the garden club could not resolve its financial difficulties in time for publication in 1932. Perhaps significantly, it was just at this time that Iden became seriously ill with what she later called "a complete knock out from a long period of over work" and removed herself from garden club activities. Wells, thinking that there was plenty of time, had suspended work on the book and had enrolled in an art school in Gloucester, Massachusetts, for a period extending until August 1. After that date, he planned to work in another such school in Provincetown until the end of the month, then meet his wife in New York on her return from a trip to Europe and make a leisurely drive back to Raleigh.[11]

It was natural, then, that when Wells learned from Couch in July that production of the book could proceed as soon as the manuscript was ready, he was taken very much by surprise. He reacted, however, with his characteristic vigor. In a rapid three-way exchange of letters, he proposed to curtail his artistic activities and return to Raleigh to resume work on the book; he then agreed to a completion deadline of 15 September, which was suggested by Tomlinson. By 11 August he was at home, working night and day at top speed. By 3 September the manuscript was in the hands of the press except for an inadvertently omitted chapter which Wells sent over in a few days.[12]

At this point, the evolution of the title, *The Natural Gardens of North Carolina*, deserves mention. The proposed work originally discussed in April 1931 by Iden and Couch bore the title "The Native Plant Communities of North Carolina and Their Habitat," a name reflecting Wells's academic ecological interest. But Tomlinson had conceived of a book on the wildflowers of the state, a rather different slant. The dual orientation that the book eventually adopted was expressed in two tentative titles mentioned by the garden club in 1931, "North Carolina Wild Flowers at Home" and "North Carolina Wild Flowers—and Their Homes." On 23 April 1931, the day of the meeting at which the garden club voted to sponsor the project, Wells called his talk to the club "The Natural Gardens of North Carolina." Iden thought the lecture was wonderful and its name "would be a fetching title for the book." Nevertheless, during the period when the book was being created, Couch, Tomlinson, and Wells consistently called its subject the North Carolina "flora," although as late as March 1932 Iden mentioned the "proposed book on North Carolina Wild Flowers." The key word "gardens" first appeared in a recorded title for the work in a preliminary announcement circulated by the garden club in July 1932: "North Carolina Wild Flowers and Their Native Gardens." It seems likely that Iden had continued to press for the title she found fetching and that Wells in the end agreed that it was an altogether natural one.[13]

In September 1932, Wells's manuscript was quickly given for review to W. C. Coker and H. L. Blomquist, professors at the University of North Carolina and Duke University, and evoked mainly technical comments from them. Couch felt that the manuscript committee of the press would accept it and asked for formal approval of the finished version from Tomlinson and Iden; they assented heartily on 14 and 16 September. By 4 October, Wells had completed some minor revisions and all was ready for the printer to set type in anticipation of a December publication date.[14]

What remained was a rush to manufacture the book in time for Christmas deliveries; that deadline was met. By 7 October type was being set by the Seeman Printery of Durham. Wells approved proofs of the cover and spine designs on 25 October. During October he also read and corrected galley proofs, the last of which he returned by 5 November. He also wrote the preface, dated 14 November, in which he credited the garden club and the university press for making the book possible, Tomlinson for providing practical and enthusiastic support, and Iden for originating the idea of garden club sponsorship. As is all too common, however, the printer experienced a delay, and actual press work did not begin until 21 November. This holdup meant that the delivery date, anticipated to be 1 December, had to be moved to 10 December. On 25 November, the day after Thanksgiving, the press sent detailed specifications for the book to the binder, L. H. Jenkins, Inc., of Richmond, Virginia; on 28 November the latter provided a sample case (cover), which was approved on 30 November. By 1 December the printed sheets had not yet reached the binder, but they arrived shortly thereafter. These events and their exact datings are described here in such detail because despite modern technological advances it seems unlikely that they would now occur so swiftly.[15]

The finished books were shipped in two lots from Richmond to the printer in Durham, that destination having been chosen because it was on a more direct motor express route than was Chapel Hill. The first batch arrived late on Saturday, 10 December, and the remainder on 14 December. It had been decided earlier that Wells would autograph the first 500 copies, the garden club subscription, so on Sunday morning, 11 December, he drove to Durham and performed that task; he finished the chore when the second lot arrived on the following Wednesday morning. Tomlinson had already provided a list of mailing addresses for 381 copies, so the long-awaited book could be shipped to the garden club recipients in ample time for Christmas, as promised.[16]

The financing of the book by the garden club moved along as swiftly as did the manufacture, although the uncertainty of the process at the time must have been painful to the participants. When Wells's final manuscript was accepted on 4 October, the press asked for the first payment of $500; Couch also incidentally expressed the anticipation that when the total amount of $1,500 had been paid, the garden club's 500 copies would either be shipped to Tomlinson or Iden or else mailed out by the press at additional garden club expense. Tomlinson paid the first $500 on 5 October but manifested surprise that the cost of shipping and handling of the 500 copies was not included in the $1,500; in any case, she definitely wanted the press to deal with distribution and not ship the whole lot to her. In reply, Couch agreed that the press would bear the postage and carrying costs after all. When Wells returned the corrected galley proofs, the press requested the second $500 on 5 November; Tomlinson sent this on 7 November. And on 14 January 1933, almost exactly thirty days after the first finished copies had been produced, she provided the press with a check for $412, all the money she had at hand; she promised the balance of $88 as soon as possible, and Couch graciously replied that it should be sent whenever convenient. A few days later Tomlinson had to present the matter to the executive board of the garden club. That group authorized the paying of the $88 from the general treasury, with the understanding that repayment would be made as soon as enough additional copies had been sold. After all the tribulations, the project had succeeded.[17]

The persons most directly involved in the struggle to produce *The Natural Gardens* were well pleased with the product of their labors. Couch was delighted with the book and stated that the project had given him genuine pleasure and satisfaction. Although Tomlinson was still wrestling with the problem of financing the work and was disappointed that the frontispiece could not have been done in color, she expressed herself as "charmed" with it and called it a "splendid contribution to Southern Horticultural literature." Proud of his accomplishment, Wells thought that the project had been carried through with a high degree of efficiency and was greatly pleased with the review of the book by Jonathan Daniels. After autographing the first group of books on a Sunday morning, that afternoon he personally delivered a copy to Iden. She was thrilled. As she held that first copy in her hands, she must have had mixed feelings when she recalled the effort to produce it: the exciting days of conception, the search for a way to publish, the difficult campaign to finance, the moments when it seemed impossible, the joyous time when everything fell into place at last, and then the crushing blow of illness that removed her from the drama as its climax neared. She extolled the book as far surpassing her dreams, referring to the "classy make up and beautiful binding"; she thought it was "a great contribution to the educational life of the state." With publication at last accomplished, all parties exchanged congratulations and probably sighs of relief.[18]

Two thousand copies of *The Natural Gardens* were printed in one press run. These had 458 text plus 20 introductory pages. The frontispiece and 209 figures consisted of 5 drawings and 225 photographs. Wells himself made the drawings and 191 of the photographs; the additional photographs included 24 by H. L. Blomquist, 7 by A. F. Roller, and 1 each by Robert Fiske Griggs, Susan Iden, and a source designated as Wooten-Moulton. But the copies of the edition were completed at three different times, and the technical specifications of finishing varied among the three lots. The 800 copies produced initially in December 1932 were bound in buckram over boards in a standard manner with sewn signatures of sixteen pages each, three tapes, green and gold silk headbands, and white endpapers. The binding bore the title on the spine and a drawing made by Wells of the Venus flytrap plant on the lower right front cover; these two items were stamped in pure gold. The spine also carried a panel with a diagram of a leaf done in dark green. All stampings were done from brass dies. The books were wrapped in jackets printed by Seeman and inserted in boxes made to fit. An additional batch of 700 copies was bound in March 1933; these were finished like the first lot but had no printed wrappers and no boxes. The last 500 copies were finished in July 1934. The bindings of these were like the earlier ones except that imitation gold was used for the stampings; these copies also had no printed jackets and no boxes but were put up in glassine wrappers.[19]

The arrangement regarding royalties for *The Natural Gardens* was rather complex. The 500 copies subscribed for by the garden club, which sold for $3.00 each, carried no royalty. The contract between Wells and the press called for a royalty to the author of 10 percent, or thirty-five cents, on each copy that sold to the public at $3.50. Wells, however, insisted that such royalties be paid to the garden club rather than to himself. A large number of copies were sold to schools for only $1.60, and no royalty was paid on these. The record in the garden club yearbooks shows that in the period from 1933 to 1939 the club received a total of $223.05 in royalties, an amount that would correspond to about 637 copies sold at the public price of $3.50. This fact suggests that as many as 863 copies from the original edition may have gone to schools.[20]

The economic history of the book can be worked out from the records of the university press. The total cost for printing was $2,352.95, including paper, illustrations, wrappers, and freight. Binding costs totaled $808.00, including folding, boxes, and dies for stamping. Handling costs to the press must be estimated from assertions made by Couch, but a reasonable figure is $700. The total cost of the 200 copies was thus about $3,900, or $1.95 per copy. From a different angle, if the $3,900 total cost was first reduced by the garden club subscription of $1,500, then the total cost to the press for 1,500 copies was $2,400; this works out to a cost per copy of $1.60, exactly the price at which books were supplied to schools. If the figures of 637 copies to the public at $3.50 and 863 copies to schools at $1.60 are used, then total sales of the 2,000 copies amounted to $5,110.30, for an average price of $2.56 per copy. Deducting $1.95, the cost per copy, from this average price yields a gross margin of 61 cents per copy or $1,220 in total. Correcting these figures for the royalties paid on 637 copies leaves a net margin to the press of $997.05 total, an average of 50 cents per copy. These figures, of course, do not take into account costs to the press of general operating expenses such as staff salaries. They are detailed here merely to suggest interesting comparisons with what present-day figures might be.[21]

In a general vein Couch later wrote, "In order for people to work at thinking and writing, they must know that if what they do is worth publishing, it will be published. They must have easy access to a publisher, the publisher must be capable of giving sound advice and criticism and must be strong enough financially to handle good material regardless of whether it can be made to pay for itself." The original edition of *The Natural Gardens of North Carolina* did ultimately pay for itself. From subsequent events it is clear that the book would have done so even without the subsidy from the garden club, but this was not evident to anyone at the time. The press imposed the condition of a club subscription because it rightly felt that the venture even then was a gamble. It is a credit to the press and to the garden club that both seized the opportunity to provide the public with a significant book and that together they carried out the project successfully. Their vision was proved correct.[22]

CHAPTER 8

Natural Gardens Revisited

Some one has mentioned that bloddy [*sic*] NATURAL GARDENS OF NORTH CAROLINA again—in a talk or on the radio or something. We've had two inquiries in rapid succession, and I foresee more. I'm tired of writing and discouraging people about this, so now I'm suggesting that they write to you and ask you to reprint it. I still think you could do real well with this book—goodness knows we get steady requests for it, and our waiting list got so long I quit putting names down.—Bunny (Mrs. Paul) Smith, 1959

THE FIVE HUNDRED autographed copies of *The Natural Gardens* which made up the garden club subscription soon became scarce. By 1933 only forty remained unsold, and by 1935 the number had dwindled to a mere thirteen. In March of the latter year Tomlinson expressed the feeling that such copies would soon be rare and lamented, "I don't believe there will be a reprint for some time." As it happened, "some time" became far longer than she probably ever imagined. Copies from the last batch bound continued to be sold until at least 1939, as shown by royalties paid to the garden club. By 1942, however, the book was out of print.[1]

The question of revising or reprinting *The Natural Gardens* was approached with good intentions on all sides. When he ordered the binding of the last five hundred copies in July 1934, Couch realized that the first edition would soon be exhausted. From as early as January 1933, his thoughts had been touching the matter of reprinting, for he then mentioned the possibility of getting out a new edition in two or three years. At that time he also envisioned expanding the coverage of the book to include the entire Southeast, not just North Carolina. This was the germ of an idea which unfortunately later bulked as an almost insurmountable obstacle to getting the book back into print. In 1934, after conferring with H. L. Blomquist, Couch wrote to Ruth M. Addoms and Henry J. Oosting, botanists at Duke University, seeking their suggestions for a possible revision covering the enlarged area. Both replied that for its intended purpose the book was excellent as it stood, although they would have liked to see the mountain and piedmont vegetation treated as extensively as that of the coastal plain. Addoms emphatically opposed enlarging the coverage, citing the likelihood that the identification keys for the larger region would become either clumsy or poorly defined. "What can be done well for a small area," she wrote, "cannot easily be done well for a larger area."[2]

Couch was giving his imagination little bridle. In March 1933 he had suggested that Wells might rewrite *The Natural Gardens* as a science reader for use in grammar schools; he thought that Mrs. Wells might join in such a project because she was a schoolteacher. Wells's response was noncommittal. Couch raised this question again in July 1934. He first observed that if summer schools continued to use the book as they had been, it could probably soon be reprinted for a cost of $2 per copy. Then he again asked Wells to consider a revision covering the southeastern states, coupling this proposal with the suggestion that it be done as a supplementary reader for the sixth and seventh grades in collaboration with Mrs. Wells. This time Wells expressed interest but thought that first he would need to examine carefully the science program in the schools to be sure the content of the book would be relevant. In any case, however, he then had no time for the project because he was working in the summer program of the New College of Columbia University and also was heavily involved in curriculum revision at North Carolina State College. So the matter of reprinting or rewriting rested.[3]

In September 1942, after the book had gone out of print, Couch again approached Wells. This time he meant business, for he carefully detailed possible royalty arrangements. His proposal for revision contained two key elements: the book should become *The Natural Gardens of the Southeast* with suitably enlarged treatment, and completely new illustrations should be provided. In reply, Wells stated that the proposal carried important implications that should be gone into thoroughly so he would like a personal conference. Couch agreed. But the rationing of World War II was heavily in force at the time, and Wells was unable to accumulate enough gasoline for even the short drive to Chapel Hill. Instead, in October he composed a detailed response to Couch's proposal. First, expansion of the coverage to the Southeast would necessitate dropping the manual half of the book because the simplified key system could not be made to work in the larger region; in this evaluation he echoed Addoms's opinion of 1934. But from what Wells could gather, *The Natural Gardens* was valued as much for its manual section as for the first half; perhaps the former part could be revised separately as a wildflower bulletin. Second, making a book out of just the first half of the original work would require a full account of the plant ecology of the region. Wells thought he had a good foundation for this, having just reviewed all the scientific work on the southeastern coastal plain in a professional publication, but a popular or semipopular account would need detailed descriptions of especially interesting areas. Even then there was a question as to the market for a popular book on the plant ecology of the Southeast. Third, although new photographs would be desirable, they would have to be made in the actual locales. In summary, Wells felt that he could indeed implement Couch's proposal in a satisfactory manner, but doing so would require a year of freedom for him to travel about the region, an impossibility under the existing wartime conditions.[4]

Couch wrote to Wells agreeing that present conditions made the work of expansion and revision impossible but suggesting that they ought to be ready to proceed as soon as the wartime restrictions were removed. Several months later, he reassured Wells on the question of demand for the projected book: he thought it would sell a few hundred copies the first year and then two to three hundred copies per year for the following ten to twenty years. His reckoning was based partly on the opinion that the ecological portion of *The Natural Gardens* was more popular than the manual section. He also thought that many of the needed photographs could be obtained from commercial sources, although it might prove possible for Wells to do them; this question could be resolved later when work had progressed. Couch also introduced a new proposal: he suggested that Wells write a college textbook in botany illustrated with plant materials characteristic of the South. Wells responded to the latter suggestion by pointing out that a botany text would deal mostly with structure and function, which lack regional connotations. Besides, he wrote, "There are entirely too many already in existence." As to the projected revision, he was now totally involved in the wartime military geography program at the college with no sign of relief and could foresee no actual work being possible until more normal conditions returned. For the duration of the war *The Natural Gardens* would have to wait.[5]

The war ended, and years passed, and yet the book remained out of print. The matter of revision was still simmering, although the situation was complicated because in 1945 Couch had moved on in his career, first directing the University of Chicago Press and later serving as editor of several prominent encyclopedias. The University of North Carolina Press was now operating with a new director and a new assistant director, Lambert Davis and Porter Cowles, respectively. In response to an inquiry in 1949 Cowles indicated that demand for the book was continuing, that talks had been held with Wells about a revision with new illustrations, and that the latter felt that his retirement as department head at State College would give him time to work on the project. But no book appeared, and in response to another inquiry in 1951 Cowles lamented that she had been trying without success to create interest in a revised edition. "Dr. Wells," she claimed, "is not really interested in the thing himself and it is hard to make any progress from there." Davis also later summarized a discussion with Wells while he was still teaching at State College, in or before 1954. Davis, like Couch before him, was convinced that the market was too small for a book about North Carolina alone. And so, although he did not insist on coverage of the entire Southeast, he proposed that the area be broadened to include at least Virginia, South Carolina, and northern Georgia. He thought that such an enlargement would capture sales in population centers such as Richmond and Atlanta. Wells found an extension feasible but did not want to undertake the large amount of work required for it because he was more interested in other things. In other words, Davis could see only a much broadened book, one on which Wells refused to work. This discussion apparently set up the deadlock that persisted for more than a decade.[6]

But demand for the original book did not abate, as shown by requests directed at booksellers, the publisher, and Wells himself. In response to such inquiries, the reasons cited by the press for a lack of reprinting varied interestingly. When the national president of the Wild Flower Preservation Society asked in November 1942 about "the very best popular presentation of the subject ever published," the war could be given as the factor preventing reprinting. Again in 1949 it was still possible to blame the previous wartime conditions. Other inquiries are on record for 1951, and in 1953 a garden club member reported that she had been forced to pass up a used copy because its price was "too, too high" at $20; "I for one," she wrote, "would like to have it reprinted, and feel it would have a good sale." Davis's response was that high printing costs for such a book prevented "putting out a new edition at a price low enough to attract a sufficient audience." Later, in 1956, the press received an order for the book "at any cost." Also in 1956, Olivia Burwell, librarian of the Greensboro Public Library, begged for a clue as to where to obtain the book because the one remaining copy in her library had disappeared. "Greensboro," she wrote, "is a city of gardeners and this title is in much demand." Cowles replied that the press had not reprinted the book because Wells was not inclined to do the considerable revision deemed necessary by gardeners and because high costs precluded a realistic price. "I guess you will just have to hope your remaining copy comes back," Cowles wrote, although she did immediately inquire and locate a copy held by a specialist in out-of-print books in Durham. On the other side, in 1955 Eunice Weeks, librarian of the Camden County High School library, wrote to tell the press that she understood *The Natural Gardens* to have "definite monetary value" and that she would like to sell her library copy "for money to buy much needed books."[7]

And so the impasse continued. Wells was not spared inquiries; he reported, for example, that he was once offered $50 for a copy of the book. In 1951 the president of the newly formed North Carolina Wild Flower Preservation Society wrote pleadingly: "There is such a demand from the wildflower public for a copy of your book. . . . What can be done about it?" In 1959 Ben W. Smith, a professor of genetics at North Carolina State College, wrote to Wells to tell him that he had seen a copy at a bookshop in Chapel Hill but had been forced to leave it because of the high price of $36; he reported that the shopkeepers said that this was the first copy they had received in all their years of operating the store. Smith pleaded with Wells that the book be reissued. Earlier in 1959 one of the owners of that same bookshop dispatched a note of complaint to Davis. Bunny (Mrs. Paul) Smith wrote that their store had received so many requests for *The Natural Gardens* that they had stopped keeping a waiting list; from now on they would direct all inquiries to the press. She thought that a real market existed for the old book and that a reprint would sell well.[8]

Smith's comments stimulated Davis to approach Wells again. He sent a copy of her note and said that it described the situation truly: there was a demand for *The Natural Gardens*. Something, he thought, ought to be done, but the question was what. Davis referred to the discussions they had held some years earlier and again proposed that the book be revised to include Virginia, South Carolina, and northern Georgia along with North Carolina. He still felt that the high production costs of the day required a larger edition, which in turn necessitated an extension of the geographical area covered. Unfortunately, Davis made a serious tactical blunder, given Wells's personality: he put forward the idea that a younger botanist be found to execute the expansion and revision, provided Wells agreed to this course of action. Wells most emphatically did not agree. He echoed opinions expressed since 1934 in pointing out that expansion to the greater area would make it impossible to construct workable simplified keys for the identification section. Furthermore, he wrote, the "suggestion of a revision by a 'younger botanist' is out of the question"; the one person he thought might have done such a job, Stephen G. Boyce, was not available. Anyway, the demand that existed was "for the book as it was published and not based on any hypothetical volume concerning the Southeast." He concluded, "You will recall that at the time of our former conference, I offered to make necessary revisions and additions to the 'North Carolina' book. These I could have easily done during the ensuing summer. This offer still stands. Otherwise, I suggest we forget the whole matter." Davis did think otherwise, for he was utterly convinced that economics dictated a larger edition and thus broader coverage. Once more the matter was dropped.[9]

A small but steady demand for *The Natural Gardens* continued, however, and copies were nowhere to be found. In November 1959 Cowles wrote in response to an inquiry, "I wish I could tell you which way to turn on trying to get a copy. . . . So far as I know it is almost an impossible situation." The demand was spurred at intervals whenever the book was mentioned in the popular press, as it periodically was. In 1958, for example, the magazine the *State* excerpted some passages dealing with the longleaf pine forests of the past. And in 1965 Bill Sharpe, writing in the same magazine, increased the pressure significantly. His article described the book in enticing terms. He also related how once when stranded near Soco Gap he had in effect traded his copy of the book for a night's lodging, only to find that the replacement cost was a prohibitive $40; the latest quotation he had was for $60. "Until this book is revised and republished," he concluded, "a large gap in our literature will go unfulfilled."[10]

Sharpe's article evoked an immediate response from Davis, for it hit him, he wrote, "on a tender spot." He outlined his view of the situation. He had been thinking about *The Natural Gardens* for many years and had held several discussions with Wells; the most recent of these had been in the summer of 1964 when the two happened to be in adjacent houses while vacationing in the South Toe River valley near the Black Mountains. According to Davis, Wells thought the book needed much revision but did not want to do the work himself; another field botanist would be needed. But there was also a new difficulty, cited now for the first time: an entirely new set of illustrations would have to be made, a job the press could not underwrite. The engravings used in the original book had disappeared, and the photographs from which they were made had disintegrated. After the book was originally published, the only other mention of the engravings in the records of the press had occurred thirty years earlier when Couch inquired apprehensively of the printer concerning their state; he wanted them to be cared for properly so there would be no danger of their warping. The printer assured him that the plates were in good condition and would be kept on file pending instructions. How and why the plates disappeared is not on the record, although Wells was later told that they had been sold as scrap metal during World War II. In any case, without them the only way of reprinting the original book was by the photo-offset process. Davis, however, judged that the result would be so substandard that no good publisher would want to use the method. "We seem," he wrote, "to have hit a stone wall."[11]

Sharpe was not put off. He immediately replied that although he understood that Wells was "sitting under his fig tree up at Wake Forest" and had no intention to undertake revision, he doubted that much revision would be necessary. He also thought that getting replacement photographs would not be a big problem because, as he put it, "a lot of suitable pictures have been made since the 30's." Davis countered that getting out a new edition would not be as easy as Sharpe seemed to think. But in connection with another work which the two were discussing, Davis admitted that there had been recent improvements in offset printing which permitted economical small editions of reasonable quality. And he promised to be more active in trying to get the book back into print.[12]

Within days, Davis approached Wells again with what he thought was a new and promising angle. His letter betrays a large excitement over the possibility of revision and a large lack of understanding of B. W. Wells. The press was preparing to publish the *Manual of the Vascular Flora of the Carolinas*, an important technical botanical work being done primarily by three botanists at the university in Chapel Hill. This project was winding down, and Davis approached one of the three about working on an enlarged revision of *The Natural Gardens* and received an encouraging response. As Davis outlined in his proposal to Wells, the revision would broaden the coverage to include South Carolina as well as North Carolina. He thought that most of the necessary fieldwork had already been done by the Chapel Hill botanist in preparing the *Manual*; the main task would be getting the necessary photographs. The other botanist was most anxious to have a definite understanding of the literary rights involved in a major revision so Davis had assured him that Wells had indicated a "willingness to turn the book over to any trustworthy younger scientist who wanted to do a proper revision." Then after an ambiguous discussion of royalties, which seemed to express doubt that there would be enough for both authors, Davis asked for a clear statement of Wells's position in the matter.[13]

He got a very clear statement. When found later among Wells's personal papers, the two-page letter from Davis consisted of eight torn fragments which had been taped back together. After he had simmered down, Wells asserted in his reply, "My answer to your inquiry relative to my book is simply, definitely, and finally, No! You are completely in error in quoting me as willing 'to turn the book over to a younger scientist.' I am sure you will recall my parting statement which I shall repeat again—'Let us forget the whole thing.'" Apparently taken aback, Davis responded by repeating what his understanding had been; also, he did, indeed, recall Wells's parting words but had taken them to refer only to the case in which the right arrangements could not be found. He continued with essentially the same proposals; he expressed disbelief that Wells would not want the book made available again under any circumstances. The demand existed, and the press was willing to accommodate any degree of association or dissociation with the project on Wells's part. The ambiguity about royalties was cleared up; Davis proposed either a 75–25 split between the new author and Wells or a lump-sum payment to the latter. He also introduced an implied threat: the copyright on the book had expired so it could now be reissued by anyone, although of course the press would not do this without Wells's consent. Wells was unmoved. The entire text of his simple reply was, "You failed to note the word 'finally' in my last letter or you would not have written to me again."[14]

Davis was perplexed. He sought the advice of Donald B. Anderson, vice-president for academic affairs of the University of North Carolina system, who from 1925 had been a friend and colleague of Wells's in the Department of Botany at North Carolina State College. Davis sent over copies of the correspondence he had exchanged with Wells and asked for an opinion on what to do. Anderson's recommendation was simple and direct: no further effort should be made to persuade Wells into a collaboration in revising *The Natural*

Gardens. The subject had come up during a recent visit with Wells, Anderson wrote, "and his emotional reaction was so strong that it was clearly inadvisable to speak further about the matter." Davis gave up. He could see no further hope for the venture along the lines he envisioned. From this time until page proofs of the reprinted book appeared on 6 October 1967 the records of the press are blank. It was an impasse.[15]

The curious fact that the copyright on *The Natural Gardens* had not been renewed introduced a new factor into the situation. A new force also entered the picture in the form of the North Carolina Wild Flower Preservation Society and in the person of its president from 1966 to 1968, Henry Roland Totten. Totten was a professor of botany at the University of North Carolina and a longtime leader in the cause of nature. He and his wife worked for many years to create garden clubs in North Carolina. He also helped in the founding of the state Wild Flower Preservation Society and was instrumental in the development of the North Carolina Botanical Garden; later a gift from the Tottens provided the latter establishment with an educational and administration building. Totten was also a close friend of Wells who fervently wished to see *The Natural Gardens* again in print.[16]

The North Carolina Wild Flower Preservation Society had long been interested in the case of *The Natural Gardens*. At its fall meeting in 1960 a resolution was passed requesting the garden club to ask Wells to revise and republish the book "or otherwise to just reprint it"; nothing came of this entreaty. Now in 1967, however, Totten's role was decisive. He held discussions on the matter with Davis, without moving him. He was able, however, to stimulate the Wild Flower Preservation Society to action, which finally led to the result so widely desired. Meeting on 5 March 1967, the executive board of the society first took note of the eighty-third anniversary of Wells's birth on that date by signing a card for him. The group then heard Totten on the possibility of simply reprinting *The Natural Gardens* without any revision; the record of Dover Publications, Inc., in reissuing out-of-print, out-of-copyright books was mentioned in this discussion. Then at another meeting of the executive board on the following 20 August real action was precipitated by Totten and Lionel Melvin, another longtime champion of nature in North Carolina. Together they proposed that the Wild Flower Preservation Society sponsor the reprinting by the university press of several hundred copies of chapter 6 of the book, "The Most Beautiful Gardens: The Grass-Sedge Bogs or Savannah Lands." (Curiously, this chapter has two titles, the one heading the chapter itself and a slightly different one in the table of contents; the same situation obtains for chapters 1, 2, and 10.) Chapter 6 includes a description of Wells's beloved Big Savannah and is perhaps the most poetical one in the book. The idea of Totten and Melvin was that the single chapter would serve as an advertisement for reprinting the entire work. As an example of the result that could be obtained by the process of photo-offset printing, Totten distributed to the group sample pages which he had secured. The Totten-Melvin proposal was accepted, and the means of raising the necessary $135 to $160 were discussed, with Herbert Heckenbleikner of the executive board offering to assist in the financing if necessary.[17]

Now progress, so lacking for so long, occurred rapidly. Totten showed his sample pages to Wells, who was delighted with their quality. On 3 September in a communication to Porter Cowles of the university press, Wells formally consented to a reprinting of the entire book by photo-offset printing with "no changes whatever from the original 1932 edition," except that "a few major new data involving corrections chiefly could be put into an addendum of a few pages." Wells prepared this addendum later in September. Lambert Davis did not acknowledge the developments until 10 October, when he wrote to Wells to confirm the arrangements and to agree formally to royalty payments of 10 percent. In his reply, Wells could not resist a final thrust: "It is interesting to note," he wrote, "that the natural world is so stabilized that books about it are never out of date."[18]

The actual reprinting was done in Raleigh at the print shop of North Carolina State University. The copies of the book used as sources for the photo-offset process were Wells's own two personal specimens from the original edition; one of these was a copy interleaved with blank pages for corrections and additions which Couch had caused to be specially made for him in 1932. Neither the two copies nor their remains were ever returned to Wells. At the fall meeting of the Wild Flower Preservation Society on 1 October the reprinting was announced and orders were solicited; the price of the book was $6.40 before 1 November and $7.75 thereafter. A draft of a press release was sent to Wells on 2 November, and the book itself appeared shortly. At the meeting of the executive board of the Wild

Flower Preservation Society on 18 February 1968, Wells was present to autograph copies of the long-awaited book.[19]

The reprinted version of *The Natural Gardens* differed only in minor ways from the original one. The binding was simple, without artwork. A new title page omitted reference to Wells's Ph.D. degree and institutional affiliation, and the table of contents was interposed between the preface and the introduction. Two paragraphs were removed from the preface, one on the value of the ecological approach and a cautionary one on the use of the identification keys. Two paragraphs were also deleted from the introduction, one on the importance of topography and soil type in affecting plant distribution and one on the terraced nature of the coastal plain. Finally, a three-page addendum explained that the wind effect on coastal plants was actually a salt-spray effect, described the hypothesis that the grassy balds of the mountains were of Indian origin, and lamented the passing of the Big Savannah.

Like the original edition, the reprinted book was greeted with favor by both popular and scientific reviewers. The *Durham Morning Herald* described its approach as geographical and the demand for it as justified by its authoritative nature. The *Charlotte Observer* characterized it as "a unique book about a unique state." The reviewer for the *State* magazine called it "invaluable" and "calculated to fascinate even the casual gardener or botanist, and yet serve as a fine handbook and reference." Earl L. Core, a professional botanist, judged it to be as accurate and valuable as when it was first issued; despite the great development of color photography and reproduction in the meantime, he found the black-and-white illustrations excellent and mostly well printed. "For the student of the North Carolina flora, and, indeed, of the plant geography of the eastern states," he wrote, "this book continues to be indispensable and the publishers are to be congratulated for having made it so easily available again."[20]

Throughout its life *The Natural Gardens* stimulated requests for the use of material from it. From the beginning the original edition had evoked such applications. The *State* magazine used excerpts in 1958. Most requests, however, were for illustrations. Violet Niles Walker, editor of *Garden Gossip*, published by the Garden Club of Virginia, in 1933 asked for and received the engravings for four figures from the book, figures 35, 36, 45, and 203, which she planned to use in the February and March 1933 issues of the publication. The Job P. Wyatt

& Sons Company considered using some plates in its seed catalog for 1935. And in 1937 R. Bruce Etheridge, director of the North Carolina Department of Conservation and Development, expressed interest in using illustrations from the book in a folder devoted to wildflowers of the state. The most popular picture was that of a Venus flytrap plant clutching a grasshopper (figure 68). Maude Wood Henry asked for the use of this illustration in 1933, as did Louise Klein Miller in 1936; the latter wanted it for her projected book, *Philosophy of Gardening*. E. V. Deans, Jr., also used the photograph in an article in *Nature Magazine*, and Max Walker deLaubenfels, a biologist specializing in sponges, employed it in his biology textbook.[21]

The most unusual application of material from the book occurred after it was reprinted. Like many works of other kinds, but perhaps uniquely for one about plants, *The Natural Gardens* served as the basis for a film. Hollis J. Rogers, a botanist at the University of North Carolina at Greensboro, was a longtime admirer of both Wells and his book. He also lamented the continuing disappearance of the natural vegetation of the state as a result of the development of the landscape by humans. Rogers conceived the idea of filming the plant communities described by Wells, with narration consisting of direct quotations from the book. In 1971 he received permission from the university press and Wells to use excerpts for this purpose. With support amounting to only about $3,000 from the university at Greensboro and the North Carolina State Parks Division, he traveled more than fifteen thousand miles around the state, filming as he went. Four years later this project culminated in a thirty-minute production bearing the same title as the book that inspired it. It was shown before diverse groups on various occasions and in March 1976 was broadcast by the university television station as part of a one-hour program. The film opens with a sunrise on the coast, moves across the state through the various plant communities, and concludes with a sunset on Mount Mitchell. The vegetation appears against the narrative from the book and background guitar music. Also included is an introductory reading by Wells himself at his farm home. The producer's goal was to create a timeless story of plants and natural vegetation; aside from Wells, only three persons appear incidentally, and care was taken to show no datable objects such as automobiles. Rogers aimed "to get on film the North Carolina that Dr. Wells was writing about before we completely destroyed it."[22]

The reprinted *Natural Gardens*, essentially identical to the original of 1932, found a small but steady market. In the first two years after reprinting, when the deficit of so many years was being satisfied, 1,017 copies were sold. In the following ten years, from the fiscal years of the university press 1970 to 1979, 1,867 more copies were sold, for a relatively steady average rate of 187 copies per year. In the seven fiscal years subsequent to Wells's death, 1980 to 1986, 509 copies were sold, an average of 73 per year. The total sales through fiscal 1986 thus amounted to 3,393 copies, significantly more than for the original edition. Only in January 1987 was the book declared out of print. Wells's work was never designed and never destined to be a best-seller in the usual sense. But in the limited market for works of its kind it has been, and after 1967 additional copies had to be struck in 1971 and 1979. If the book had been reissued years earlier, one can only guess how many more copies might have been sold; if it had been reissued when it might have been, one can only surmise how many more readers of the generation between printings might have enjoyed the instruction and pleasure it delivers.[23]

The Natural Gardens of North Carolina is a unique book, as its author was a unique man. It was conceived by a scientific mind endowed with the rare ability to communicate technical knowledge in simple and fascinating terms. It was nurtured by lovers of plants who recognized the genius of a true teacher. In the face of seeming economic impossibility it was resurrected by the force of public demand, an insistence aroused by its timeless strength. In its field the book became a classic that refused to die.

PART 3

Academician

CHAPTER 9

Professor and Head

I shall never forget my impressions, coming from a well-equipped laboratory and a big university. . . . It was just a shock. . . . In my first class boys came in overalls, chewing tobacco, and they were not used to the kind of course I was giving, dealing with theory and with molecules and forces and physics. . . . Many of the students used to keep guns in their dormitory rooms and fire them off at night for excitement. . . . The institution was totally oriented toward pragmatic ends. The School of Engineering was teaching boys how to do surveying and laying out sewer lines and things of that sort. The School of Textiles was training mill superintendents. The School of Agriculture, the experiment station, was running line-row fertilizer tests. . . . There was no interest in basic theory or what we would call today fundamental research, . . . in the administration very little appreciation of what a university really ought to be. It was in the early stages of development. . . . [But] the botanists were a lively group. . . . I was fortunate in the department I was in.—Donald Benton Anderson, 1978

WELLS WAS AN academic scientist; every professional position he ever held was on the faculty of an institution of higher learning. For four years after earning the B.A. degree and later for two years after receiving the M.A. and Ph.D. degrees he taught at a total of five schools in several regions of the country. But then he joined the college at which he spent his remaining thirty-five active years. To a large extent, the course of his life was determined by his casting his lot where he did. Reciprocally, the nature of that institution was partly determined by him, for he contributed significantly to the shaping of what has become North Carolina State University.

In the fall of 1919, at the age of thirty-five, Wells became professor and head of the Department of Botany and Plant Pathology at the North Carolina State College of Agriculture and Engineering in Raleigh. He was appointed on the strong recommendation of officials at the Kansas State Agricultural College, where he had earlier taught for two years before beginning graduate study. To interview Wells before hiring him, President Wallace C. Riddick sent Dean of Agriculture Charles B. Williams out to Arkansas, where he was then employed. The outcome of the interview was successful from Wells's point of view. He was dissatisfied with his position in Arkansas and would gladly have come to Raleigh for less money than was offered.[1]

The department he joined was a primitive one. There were three faculty members, including himself, but the other two were plant pathologists, interested in diseases of plants rather than in the plants themselves and working primarily on practical matters of disease control with the agricultural experiment station. In effect, Wells was the only botanist and was expected to do almost all of the botanical teaching in the college. Patterson Hall, the building in which the department was housed, was a glorified barn. Although the exterior was made of brick, the wooden interior had never been properly finished: it consisted of a frame of foot-square posts and beams with floors fashioned from planks laid across. These planks abutted only loosely, and every day each floor rained dust, seeds, and plant debris into the offices and laboratories below. Two large glass panes made up each of the poorly fitting windows; almost every thunderstorm would break several of them, and every winter found snow sifting through onto tables, desks, and floors. The available equipment matched the building: it was little in quantity and poor in quality. Although years later the department occupied somewhat better quarters in Winston Hall, it did not have adequate facilities during Wells's entire term as head.[2]

When he arrived, Wells quickly exerted efforts to put arrangements on as sound a footing as possible. He organized the facilities, which he termed "chaotic" as he found them, in a rational way and "threw out mounds of junk." He was able to induce the college administration to buy some new microscopes and other equipment for instructional use, and he designed new hardwood tables for microscopic work. The latter were wedge-shaped so that each student could make the best use of skylight from the windows because electric illuminators were not provided; even so, the use of microscopes on overcast winter days was difficult at best. He managed to convince President Riddick that an additional faculty member was needed at once to handle the instructional load. Hired for the purpose during Wells's first semester in 1919 was Ivan Vaughan Detweiler Shunk, who for years afterward taught a full complement of botany courses and was solely responsible for the work in microbiology.[3]

An even more serious problem for Wells than the poverty of the department's resources was the general intellectual atmosphere of the college. Known until two years previously as the North Carolina State College of Agriculture and Mechanic Arts ("the A. and M. College"), the institution when he joined it was a small, struggling technical school with no tradition of scholarship. It had so far been unable to achieve accreditation by the Association of Colleges and Secondary Schools of the Southern States. The college buildings were shabby and antiquated, sited somewhat ignominiously along the railroad tracks of west Raleigh. The 1,049 students were thought by outsiders to be culturally raw and lacking in social grace. Of the sixty-eight academic faculty only six, including Wells, held the Ph.D. degree; thirty-three had earned only the bachelor's degree, and seven had no academic degree at all. The library contained a total of only 9,830 volumes and received only 335 periodicals. The college still had the reputation of a trade school, for it was totally oriented toward practical ends; there was little interest in basic theory and little appreciation of what a university, or even a college, ought to be. Faced by these characteristics, on assuming office in 1923 President E. C. Brooks made it his goal to improve drastically the academic quality, professional status, and physical plant of the institution.[4]

B. W. Wells at work in his office at North Carolina State College, spring 1939. Courtesy of Larry A. Whitford.

Although Brooks's efforts resulted in some progress, the situation was not dramatically improved during Wells's first decade at the institution. Evidence for this fact was the report of a committee of outside educational experts in 1932, which recommended transfer of State College to the University of North Carolina at Chapel Hill; it proposed that the buildings and grounds be used by the state or by the city of Raleigh, possibly for a junior college. The committee described the quality of the college as follows: "The present nucleus of organization and personnel at Raleigh, while relatively efficient for its present sphere of work, scarcely affords the foundation for a scientific school of notable rank. The staff includes few men of the first rank, either in point of training or of professional attainments. The scientific equipment is inadequate for work of an advanced grade." Even by 1932 Wells had certainly demonstrated that he was one of those "few men of the first rank," but it was not without struggle.[5]

From the moment he arrived, Wells by precept and example was a proponent of academic excellence. In 1920 he and entomologist Zeno Payne Metcalf, a colleague and neighbor who became his lifelong friend, organized a faculty research club. The purpose of this organization was to encourage the faculty as a whole to pursue research, publish in scholarly journals, participate in learned societies, and attend national meetings. The group was enthusiastic and effective enough that the trustees of the college were persuaded to establish a modest fund for the support of faculty research. And in 1932 and 1933, partly in response to a summary dismissal of a scholarly dean of the graduate school, Wells and Metcalf were instrumental in the founding of a State College chapter of the American Association of University Professors (AAUP). They were joined in this venture by twenty-seven other members of the faculty, including the botanists I. V. D. Shunk and D. B. Anderson. In addition to the general objectives of the national organization, this group aimed to promote excellence in teaching at the institution. The research club and the AAUP chapter are two examples of how for many years Wells and Metcalf led the botany and zoology departments as continuing forces for the ideals of a university.[6]

From the beginning Wells was determined to expand the botany department as needed to ensure high-quality instruction. In addition to Shunk, he was able to add Alexander Campbell Martin to the staff in 1921. Martin obtained the M.S. degree under Wells in 1924 and after leaving State College in 1926 eventually became a recognized expert on seed morphology and wildlife food with the United States Fish and Wildlife Service; he was replaced by Larry Alston Whitford. But Wells was not interested merely in adding numbers to the staff: he insisted on hiring the best faculty obtainable. When he convinced the administration in 1924 that a plant physiologist was needed, he screened candidates from throughout the country and eventually hired a student from the Johns Hopkins University laboratory of Burton E. Livingston, one of the leading physiologists of the day. This man came highly praised and recommended, but he turned out to fall far below Wells's expected academic standards. Wells dismissed him after only one year and was furious that Livingston had thought mediocrity to be good enough for his department. He was more careful in choosing a replacement in 1925: he personally traveled to Columbus, Ohio, to interview Donald Benton Anderson, and when the latter arrived in Raleigh he and Shunk in effect subjected the young man to an oral examination on his botanical knowledge. This time the result was a happy one. Anderson for many years was a leading member of the State College faculty and eventually succeeded Wells as department head.[7]

Wells continually urged his staff to achieve their best in scholarship. He encouraged and helped Shunk and Whitford to obtain Ph.D. degrees and tried to stimulate all to carry on research and be active members of their profession. He provided more than encouragement: he set a striking example. His own teaching was excellent and his research productive, and he participated in professional societies. He was elected a fellow of the Ohio Academy of Science in 1921, a fellow of the American Association for the Advancement of Science in 1925, and president of the North Carolina Academy of Science in 1933. He served as vice-president of the Southern Appalachian Botanical Club in 1939 and as president from 1943 to 1946. Over the years he was a member of the Ecological Society of America, the Botanical Society of America, the American Society of Plant Taxonomists, the Torrey Botanical Club, the Association of Southeastern Biologists, the Entomological Society of America, the Eugenics Society of America, and the Society of the Sigma Xi. Late in his career he served as an associate editor of the scientific journal *Ecology* and as a consultant to the National Science Foundation.[8]

During the thirty years of Wells's headship, the botanists of his department increased in number from one to eight, almost all of whom became nationally prominent. They included Wells, Shunk, Anderson, and Whitford and later Murray Fife Buell; after 1946 Buell left, but added were Ernest Aubrey Ball, William Basil Fox, Robert Kenneth Godfrey, and Herbert Temple Scofield. (The story of the plant pathologists in the department is a separate one, not considered here.) More important than the number of botanists were the cohesiveness, vigor, and enthusiasm created in them by Wells's intellectual stimulation. Wells and Metcalf arranged joint seminars of botanists, zoologists, and entomologists at which faculty members not only presented papers but also discussed them critically as a group. These were lively and stimulating sessions, which in later years were recalled by the participants with great pleasure. Especially in the period between the two world wars, intellectual life among the botanists at North Carolina State College was remarkably interesting and challenging, and Wells was the catalyst that made it so.[9]

Wells had no patience with poor scholarship and would not tolerate intellectual dishonesty. His response to an event involving the latter is revealing. At one of the joint seminars in about 1924 an entomologist from the state Department of Agriculture, who had been attending regularly, delivered a paper outlining what he put forward as his own ideas about some possible future developments in science. In discussing the presentation later, the botanists Martin and Shunk found that the thoughts expressed were exactly those contained in a recent article in the *Century Magazine* by the eminent biologist J. B. S. Haldane. When given this information, Wells was livid with rage; he prepared an accusatory statement which he read publicly at the next seminar. After providing opportunity for response and receiving only a transparently weak defense, Wells ordered the offender never to attend the seminars again. He never did.[10]

When he assumed the presidency of North Carolina State College in 1923, Eugene C. Brooks launched a program through which he hoped to improve all aspects of the institution. At one of the first meetings with the faculty, he stated, "We *must raise* the academic level of this institution." A local newspaper reported Brooks's reorganization of the college structure with subheadings that included the following words: "Best thought must direct," "Instruction will be linked with research," and "Dead wood, though, must go." Wells was wholeheartedly sympathetic with the announced sentiments and intentions of Brooks's program. He, along with Metcalf, was supportive in 1925, when Brooks brought about the resignation of Dean Benjamin Wesley Kilgore of the School of Agriculture largely because of the state of the teaching program in that school; he took this position even though it also led to the departure of the plant pathologist Frederick A. Wolf, an able colleague who would not side with Brooks against Kilgore. But Wells did not allow his firm commitment to academic excellence to be tempered much by political expediency. Brooks's administration was marred by a drawn-out series of events which eventually led to the dismissal in 1931 of the outstanding dean of the graduate school, Carl C. Taylor, "a liberal sociologist with a knack for irritating reactionaries." In this face-off Wells was on the side of Taylor even though he knew that it would lessen his influence with the administration. His colleague L. A. Whitford later recalled that he "fought so vigorously for progress and the improvement of scholarship at the college that he was at times unpopular. If he believed in it, he espoused an unpopular cause as readily as a popular one."[11]

In administering his department Wells was an effective head who never aspired to hold a deanship or other higher academic office.

He disliked detailed paperwork but dealt with it promptly and so managed well the more routine aspects of administration; he was extremely proud when Chancellor John W. Harrelson characterized his department as well run. Occasionally, however, his sense of duty was overridden by the pettiness of some tasks. For example, he objected strongly to the daily written reports of student absence and tardiness, which the college administration at one time demanded, and refused to return them regularly even in the face of personal telephone calls from President Brooks. He approached organizational matters rationally, not politically. He presented departmental needs to his superiors in a logical and succinct way and expected objective consideration and fair treatment, even though he did not always receive them; only in extreme cases was he insistent. His effectiveness resulted not from scheming or skill in maneuver but from genuine leadership of his faculty. He was open to suggestion but usually took time for consideration before accepting it. He would occasionally react with a burst of vigorous refusal, but this would often be followed by a more thoughtful assent. His colleagues learned to sense when it was opportune to approach him with requests. And on crucial issues he possessed the important ability to be decisive or indecisive, as necessary.[12]

Wells favored the revolutionary change in State College which began when it was consolidated in 1931 into the University of North Carolina system, along with the university at Chapel Hill and the College for Women at Greensboro. He correctly saw that this structural change would in the long run improve the standards of the college and beneficially expand its horizon. He himself had realized the advantages of a united university and had written to Governor O. Max Gardner suggesting this idea shortly before the latter in 1930 called upon the legislature to effect the union. Wells always thought that his suggestion perhaps opened the governor's thinking to the possibility, but in fact the proposal had been made earlier publicly by others, most prominently by Josephus Daniels in 1919 and by Governor Cameron Morrison in 1922. In fact, Gardner himself had favored consolidation even before becoming governor. Nevertheless, in later years Wells frequently alluded not only to his letter to the governor but also to a secret meeting with Gardner, which he, Metcalf, and one other State College faculty member attended. The latter claim is supported by the fact that before finally pressing for consolidation Gardner privately took soundings from faculty members of the concerned institutions.[13]

Wells participated actively and wholeheartedly in the implementation of university consolidation. It was the policy of the president of the new system, Frank Porter Graham, to move slowly and carefully in effecting change and to approach restructuring through studies by faculty committees. Chancellor Carey H. Bostian later testified that Wells was "often far ahead of other members of the faculty in support of high standards of scholarship and in making a critical evaluation of courses and curricula for their role in developing basic knowledge." During consolidation, this sound dedication was recognized by Wells's appointment as chair of the important committee to examine the curriculum of State College. This committee, consisting of eleven prominent faculty members, met frequently during the 1933–34 school year. Wells's work with it was a major effort; indeed, he cited it as one of the reasons why he could not consider at that time a revision of his book, *The Natural Gardens of North Carolina*. The eighty-four-page report of the committee, submitted on 14 April 1934, contained detailed specifications for all curricula in each of the four schools of the reconstituted college (Agriculture, Engineering, Science, and Textiles). It also contained statements of principle which reflected Wells's personal views of what the academic world of State College ought to be like. For example, technical specialization was left to the final two years of a student's program, while courses of a broadening nature were introduced into the first two years of all curricula. The aim, asserted the report, was "protecting the student from being cheated out of certain fundamental and cultural contacts without which he can never be regarded as reasonably well educated." In several sections the report emphasized "high cultural values" and the idea that "education should include preparation for more complete living"; it also stated that "a sound basis for technical training includes thorough training in mathematics, physical science, and biological science." The findings of this committee clearly pointed the way for the evolution of the practical technical school, which State College had been, into the technological university it would become.[14]

The period of university consolidation was a time of both excitement and apprehension on the part of the faculty of State College, but it was also a time of enthusiasm for improvement. The botanists as a group, with Wells at their head, contributed significantly to the changes that took place and were caught up in the spirit of progress. At one point Wells and Anderson suggested that their department be merged administratively with the Department of Botany of the university at Chapel Hill. They saw this as a means of maximizing the effectiveness of both departments, which were already in many ways complementary. They intended for John N. Couch, chair of the Chapel Hill department, to assume direction of the combined unit. This proposal received little support, however, especially because Couch himself was not in favor of it, and so the union did not occur.[15]

The changes in State College following consolidation were significant. Most important was the new university atmosphere that began to pervade the institution. But developments in resources and facilities were real also. For example, in the first decade following the reorganization the number of members of the teaching faculty holding the Ph.D. degree increased from thirty to eighty-eight, from about 20 percent of the total to about 40 percent. In the library the number of volumes grew from about 33,000 to about 70,000, and the number of periodicals received rose from 515 to 886. And in 1939 the botanists finally could hope for new physical arrangements. Plans were put forward for remodeling Patterson Hall, and the department moved into supposedly temporary quarters in Winston Hall, an old engineering building, dreaming of new offices, modern laboratories, and adequate storage space. The dreams were long in realization, however, for the department remained in Winston Hall until the first unit of Gardner Hall was completed in 1952. Nevertheless, improvements were either actually made or projected, rather remarkably so in view of the severe economic depression that gripped the country at the time.[16]

That depression worked heavy hardship on the state, and North Carolina State College struggled for survival. The economic impact of those times is illustrated by the specific case of Wells's salary. In 1925, following a general increase for the faculty, Wells was paid $4,500 per year, the maximum rate for a full professor and department head; at that time the president received $10,000 and deans $6,000. But when state revenues dropped precipitously after 1929, faculty salaries were reduced drastically and then over the years raised only slowly. Calculations based on published rates indicate that after three cuts Wells's salary from 1932 to 1935 was at a low of $3,060; after three subsequent increases it was still only about $4,200 in 1937. A "war bonus" was added in 1943, but it was not until 1945 that a slight increment in this bonus brought his pay to a higher dollar value than that of 1925, about $4,615. In other words, after receiving a salary increase at the age of forty he did not see a greater pay until he was sixty-one and during most of the intervening period was earning considerably less. Of course, during these hard times many people were much worse off than Wells was. After World War II salaries were still low at State College, as Wells's reflects. His yearly earnings until his retirement were as follows: 1946–47, $5,000; 1947–48, $6,000; 1948–51, $7,200; 1951–53, $7,380; and 1953–54, $8,118. In view of Wells's accomplishments during his career, it is sobering to reflect upon what the state and the college were getting for the money paid him. Certainly his salary was not proportional to his dedication to his work and his loyalty to his institution.[17]

Following closely upon the years of mixed hardship and excitement caused by economic depression and university consolidation, the time of World War II brought disruption. With all able-bodied young males needed for military service or war industry, technical colleges for men were hit hard indeed. North Carolina State College, like many other institutions of higher learning, was saved from closure by several military programs that used its facilities and faculty in training soldiers and sailors. Wells participated in both army and army air force programs between early 1943 and late 1945. The specified curriculum of these programs did not include botany or biology, which were regarded as nonessential. Therefore, seventeen faculty members in botany, zoology, entomology, and economics were organized into a group to teach the required course in geography. A professional geographer was engaged to lead this effort, which was guided by a nationally prepared course syllabus. The academic standards of this geographer proved far too low to suit Wells, who eventually lost his temper and arranged to have him replaced by the botanist D. B. Anderson; subsequently the course was a stimulating one for both faculty and students. It was remarkably broad, including such topics as cartography, meteorology, governmental structures of various countries, religions of the world, races of mankind, and even the origin of the universe. The faculty worked very hard to master the subject matter and to teach it well. In addition to hard study of the technical material, they labored many nights to create glass lantern slides of maps, charts, and other visual aids. Wells cited his almost total involvement in this effort as the reason why he could not contemplate revising *The Natural Gardens of North Carolina* until after the war. That the course was successful was shown by the high rankings of State College students on the standardized examinations from Washington.[18]

The army air force program of which the geography course was a part was ostensibly a training effort, but its real purpose was to protect a pool of potential cadets from being siphoned off by other branches of the military. Later historians characterized the program as "a ruse which was readily detected by the Selective Service System and the War Manpower Commission. The college program was never successful in its training function." At North Carolina State College, however, the botanists accepted it at face value; they worked sincerely to make the geography course an effective educational experience. Years later Anderson recalled that "it was the botanists who gave real life and vigor to the program," and this was largely owing to Wells, whose enthusiasm was both "enormous and infectious."[19]

After the war ended, disruption of college life continued, but this time the cause was not a dearth of students but a superabundance. The great postwar influx of military veterans taxed the facilities of the school, almost swamping it, and Wells's department was as stressed as any. The already crowded quarters in Winston Hall could not accommodate the press, and for several years overflow classes in botany were conducted in temporary Quonset huts erected behind the building. These small prefabricated structures were primitive teaching facilities: they had electricity and oil heaters, but all water had to be carried in by hand. In the late 1930s the lack of sufficient good microscopes had forced the botanists to curtail laboratory microscopic work by individual students; instead, they handled much of this kind of study with projected lantern slides of appropriate structures. In fact, necessity had been turned to advantage. The department had organized the laboratory periods into group discussions led by the instructor, and as it turned out this approach had seemed to the faculty to be more effective than previous traditional procedures. In 1938 the practice had received national attention from the Botanical Society of America as a teaching innovation. Now with the overwhelming enrollments of the postwar period this method of group discussion from lantern slides as well as from actual plant materials was expanded to encompass the entire course, not just the laboratory portion. The teaching of botany became a Socratic operation, which, in the hands of the right instructors, led to effective instruction despite the scarcity of equipment.[20]

The greatly increased student enrollments necessitated the hiring of more faculty members, and so the botanists, who numbered five before the war, suddenly were eight in 1947. The pressure of these times not only disrupted the academic regimen for Wells and his faculty, it also disturbed the closely knit spirit and vigor that had existed earlier. This was natural enough, for of the eight botanists four were new, a proportion too great for instant assimilation; the development of a new cohesiveness would have taken considerable time. Unfortunately, there was not time, for Wells was compelled by regulation to relinquish his headship in 1949 at the age of sixty-five. On this occasion, the dean of the School of Agriculture wished to terminate his connection with the college, but his longtime colleague and successor Anderson would not hear of such a thing. A compromise was struck. Wells was given two alternatives: either to continue on a year-to-year basis, with a review of his performance and capabilities to be conducted each year; or to take a five-year contract, with the proviso that if he died during this period all of his retirement benefits would be lost. Characteristically, Wells was incensed at the idea of being reviewed and evaluated annually. He chose the second option.[21]

After 1949 Wells remained active as a faculty member. He pursued his research on the Carolina Bays, guided his graduate student S. G. Boyce, and taught as effectively as ever. The period following his stepping down from the headship was turbulent for his department. During this time, two faculty members died. In 1951 one of these was Wells's friend and co-worker of thirty-two years, I. V. D. Shunk; in 1952 the other was the young William B. Fox, tragically killed by his own gun apparently in the hands of his infant son. Even more new faces appeared. But new personnel, new leadership, and new conditions were accompanied by new problems and the need for new procedures and new directions. Wells's time had passed, along with his administrative authority, although his presence continued to be felt until his complete retirement in 1954.[22]

As professor and head of his department, Wells contributed importantly to North Carolina State College. Throughout his long career at the institution he consistently held to the ideals of the academy and struggled with his full force for their realization. He embodied the best that can be expected of an academician as scientist, scholar, teacher, administrator, and partisan of intellectual principle. Although many persons helped to lay the foundation for the present North Carolina State University, Wells in his sphere was as significant as anyone.

CHAPTER 10

Teacher

I enjoyed many times with him, such as walking up and down . . . , the two of us, on the sand when there were no houses back behind the dunes. . . . Both of us would go swimming and then sit on the beach or in the surf and talk about what I should be doing to do my dissertation. Dr. Wells was responsible for directing my work, but he never *directed* it. And after awhile I learned what he was really doing. . . . What he attempted to do was to develop a scholar. . . . We talked about the physics of bursting bubbles and wind movement. We talked about how plants grew and physiology. But interspersed in all this were politics, human nature, people, everything you can name. . . . We slept out on cots at White Lake and Black Lake and we argued, and discussed, and disagreed, and finally agreed on how those lakes were formed, why some are black, why some are white. . . . In later years after I left Raleigh and as I corresponded with him, the thing that kept coming back . . . always was the encouragement of scholarship.
—Stephen Gaddy Boyce, 1978

WELLS WAS consistently, enthusiastically, and effectively a teacher. Although other duties were also involved, the professional positions he held during his lifetime called primarily for the instruction of students. Almost no records exist which bear on his teaching activities early in his career, but his teaching performance at the Kansas State Agricultural College brought him strong recommendations. And Thomas L. Quay, who came to North Carolina State College as a graduate student in 1938 from the University of Arkansas, recalled that even after twenty years a faculty member at the latter institution recalled him to be "not only a good botanist and a first-class fellow, but a real live wire." But it was during his many years in Raleigh that his abilities as a teacher reached their full expression.[1]

Wells taught as a professor at North Carolina State College for thirty-five years. When he first arrived there in 1919, he was the principal teaching member of a three-man department, the other two being mainly researchers. The magnitude of the teaching responsibility in botany was enormous for one man, for after World War I the student enrollment at the college had almost doubled, from 552 in 1917–18 to 1,049 in 1919–20. Although I. V. D. Shunk joined Wells in November 1919, the task was still great. That first year Wells taught courses in plant diseases, systematic botany (identification and classification of plants), and poisonous plants; Shunk handled agricultural bacteriology and advanced bacteriology; and the two collaborated in general botany, plant physiology (processes of plants), and short courses in plant life and crop diseases. In size and breadth these teaching responsibilities would be unthinkable at a university today.[2]

The courses that constituted Wells's most creative contribution to teaching were those in ecology, the area of his major scientific interest. During his career at State College, he taught introductory plant ecology for thirty-four years. This course consisted of lectures and numerous field trips; he never incorporated indoor laboratory periods into its structure. Some idea of his approach to the subject may be gained from the lengthy catalog description he used from 1920 to 1932: "A lecture and field course presenting the basic facts concerning the influence of environment in controlling plant distribution. After a brief survey of the main vegetational areas of the world, emphasizing the United States, an intensive study of North Carolina conditions is made. Some attention is given to those structural adaptations in plants which are found associated with particular environments. The course closes with an investigation into the contribution that ecology makes to the solution of certain crop problems, especially those that arise out of soil and climatic situations." As this description suggests, a significant component of the course was plant geography, vegetation types, and their distribution on the earth. After 1932 he reduced this geographical aspect as well as the overall description: "Environmental control of plant distribution with emphasis on the habitats and vegetation of North Carolina."[3]

From 1923 Wells also taught advanced plant ecology for thirty-one years. Before 1932 this course consisted largely of individual study by the student on some aspect of the ecology of the southeastern United States. It required "a large amount of field work," "extensive readings," and "frequent consultations with the instructor." In 1932 the description of the course became "Minor investigations in vegetation-habitat problems accompanied by advanced reference reading." From 1934 on, he described it as follows: "Practice in the use of the instruments necessary in the study of environmental factors. Advanced readings and conferences on plant distribution in relation to these factors." From 1945 to 1948, he also offered an ecologically oriented course in crop geography, which dealt with the "history, distribution, and ecology of cultivated plants."[4]

In addition to his offerings in plant ecology, Wells also taught other courses, either alone or with colleagues, in some cases over long periods. At one time or another he single-handedly taught the following subjects: systematic botany, three years; advanced systematic botany, twelve years; systematic botany of grasses, eight years; microtechnique (the preparation of thin slices for microscopic examination), five years; plant diseases, four years; and medical botany, three years. He also collaborated with other instructors in the following courses: plant morphology, twenty-six years; systematic botany, twenty-four years; advanced systematic botany, twelve years; dendrology (the classification and identification of trees), twelve years; plant physiology, four years; genetics, three years; plant nature study for schoolteachers, three years; and water biology for sanitary engineers, one year. He also participated in the introductory elementary botany course, heavily at first, later only slightly, and eventually not at all. Since Wells's day the explosion of scientific knowledge has made such a teaching array impossible, or least impractical, for any one botanist to manage. But even in his time only a very broadly

A lifelong lover of nature, children, and animals, B. W. Wells even in retirement seized every opportunity to teach somebody something, about 1956. Courtesy of Maude Barnes Wells.

knowledgeable professional could handle such breadth. Wells could, and he did it well.[5]

When Wells first came to North Carolina, no state institution of higher learning had a curriculum in forestry; students at State College who were interested in this practical subject had to improvise a program in agriculture or go elsewhere. From the beginning Wells supported the idea of adding education in forestry to the college offerings; indeed, he lamented the situation publicly in an article. When the teaching of forestry finally began at the college in 1929, the botanists Wells, Shunk, and D. B. Anderson became members of the forestry faculty. This arrangement was logical because five courses in botany were required in the forestry curriculum; they included Wells's course in plant ecology and two others in which he participated with Shunk, systematic botany and dendrology. Especially through his ecology course he contributed importantly to the basic science education of all students in forestry, in addition to that of a great many in agriculture. His impact was long-lasting. In 1962 S. G. Boyce, his former graduate student, who pursued a career in forestry, testified that Wells was often "prominently mentioned by forestry graduates at state and national meetings." Forty-four years after his graduation in 1934, A. Bigler Crow, professor of forestry at Louisiana State University, vividly remembered Wells and wrote to the Department of Botany to express appreciation for his ecological teachings. And in 1983 another forestry graduate, Charles M. Matthews, testified that Wells had "made him at least as interested in ecology and environmental science as he ever was in forestry."[6]

In the classroom Wells was an unusually effective teacher. He was a dynamic and engrossing lecturer and very popular with students, not because he diluted his subjects to pap but because he challenged them with the penetrating thoughts of an able mind. In the fall of 1938, for example, when the number of students at the college totaled 2,297, some 80 of them enrolled in his course in plant ecology. In 1974 a former student, William Garland Woltz, recalled that he was "a stimulating, enthusiastic teacher, who was thoroughly enjoyed by his students. His enthusiasm was contagious—his students caught it soon after they entered his classes." But it was not merely excellent presentation that marked Wells as a master teacher; it was also his viewpoint. Although he knew the importance of facts, he did not consider education to be the mindless acquisition of detail; he wanted students to learn to exercise their rational powers. As Boyce put it, "The one point that always stands out is that he taught students to think, to observe nature, and to draw conclusions rather than rote memory." In challenging them to use their minds fully, he could lead most of them to see the power of science and the beauty of botany.

In a few of them he could even awaken the compulsion to pursue botany or biology as a career.[7]

Wells was a dynamic and effective teacher in the classroom and in the laboratory, but he was always in superlative form in the field, forcing his students to confront nature head-on. His field trips in ecology acquired legendary status, and the annual class trip through the coastal plain to the sea became an institution, especially among the forestry students. Like his extracurricular field trips, these class excursions usually involved visits to scenes of his professional research, especially the Big Savannah, and he also usually included an overnight stay at the beach complete with festive partying. Over the years between 1934 and 1954 the student publication in forestry, *Pinetum*, regularly carried articles about these trips. The titles of some of them convey an impression of the spirit the trips must have engendered: "An Ecological Fantasy," "Ecological Escapade," "Down to the Sea," "Ecology for Moderns," "Hitting the Pamlico Trail or Dr. Wells, I Presume," and "With Wells Along the Waccamaw." How students felt about Wells is clear from statements contained in these articles: "He's the most cheerful and exuberant personality imaginable, and his exuberance is contagious"; "Dr. Wells exhibits much more energy and enthusiasm than many of his more fuzzy-cheeked associates in the faculty."[8]

In addition to his capabilities as a teacher of undergraduates, Wells was a talented mentor of graduate students. Because of institutional circumstances that delayed the blooming of the graduate school of North Carolina State College until about 1950, he was able to have only a few graduate students, eight for the M.S. degree and one for the Ph.D. Because he had so few professional students, Wells never accrued the sort of "school" or network typical of the recognized plant ecologists of his day. His nine students were as follows, listed with their dates of graduation and titles of their theses:

1. Alexander Campbell Martin, 1924, M.S. in Biology, An Ontogenetic Study of the Gall *Phylloxera caryaeseptem* Shimer.

2. Carlos Frost Williams, 1924, M.S. in Biology, The Morphological Ecology of Savannah Plants. I. *Campulosus aromaticus* (Walt.) Scribn.

3. Larry Alston Whitford, 1929, M.S. in Botany, The Algae of Lake Raleigh: An Ecological Study.

4. Jesse Hickman Roller, 1931, M.S. in Botany, Ecological Aspects of the Sandhills.

5. W. Melvin Crafton, 1932, M.S. in Botany, The Old Field Prisere: An Ecological Study.

6. Robert Kenneth Godfrey, 1938, M.S. in Plant Ecology, The Compositae of Wake County, North Carolina, in Time and Space.

7. Joseph Patrick McMenamin, 1940, M.S. in Plant Ecology, The Leaf Anatomy of Southeastern Shrub-bog Plants.

8. Edward Dan Cappell, 1953, M.S. in Systematic Botany, The Genus *Scirpus* in North Carolina.

9. Stephen Gaddy Boyce, 1953, Ph.D. in Plant Ecology, The Salt Spray Community.

Five of these students, a respectable proportion, later distinguished themselves as professional scientists, Martin in plant-wildlife relations, Williams in horticultural breeding, Whitford in phycology (the science of algae), Godfrey in systematic botany, and Boyce in ecological forestry. To this inventory may be added the name of George Frederik Papenfuss, whom Wells inspired as an undergraduate to embark on a career in marine phycology and whom he regularly counseled during graduate education at another institution.[9]

Wells was an outstanding academic adviser of advanced students. Whitford reported that "he could come up with a dozen ideas for a graduate student to work on every day, almost, just over and over again." Yet he did not manage or direct his students in a detailed manner or use them as laborers in his own research. Rather, he aided and stimulated them to become scholars in their own rights through their own abilities. T. L. Quay, of whose graduate advisory committee Wells was a member, said, "I never felt I was being tested or examined, but just helped and encouraged." Wells treated students as apprentices in continuous collegial dialogues. And in honest private discussions they sensed that they were at the feet of a true counselor and responded with the best of which they were capable. Most of the advisory relationships which Wells established in this way persisted for years, even for life. In correspondence over a long period Boyce was repeatedly impressed by Wells's constant encouragement toward scholarship. Papenfuss reported continuing to seek advice long after leaving State College and never being disappointed: "He was one of those men who made students feel that they could go to him for counsel and encouragement and they always received it in full measure." And Godfrey testified, "He has ever remained a mentor without whose help and wisdom I should often have been left floundering. This I know to be true in respect to many persons." Wells's graduate students were not numerous, but they were grateful and loyal, and for good reason.[10]

In addition to his regular teaching of classes, Wells also participated in other activities involving students. A few examples will illustrate that he was not a professor who disappeared from sight when the classroom bell rang. In the 1920s he and his friend Metcalf were joint sponsors of the undergraduate biology club; during the same period he was one of the faculty advisers to the publication produced by agriculture students. In 1933 he gave a talk to all freshman students in agriculture on study means and habits. In it he asserted that the key to academic achievement is habitual concentration; he recommended daily note-taking, outlining, and self-testing. His interest in students extended beyond the bounds of State College. One of his activities in the North Carolina Academy of Science was membership in the lectureship unit of the publicity committee; in this capacity he visited high schools to deliver talks on science. Finally, although he usually spent his time outside the academic year in pursuit of ecological research in the field, he did some summer teaching. Twice he taught at the Mountain Lake Biological Station of the University of Virginia; in 1939 his course was entitled "Plant Habitats," and in 1950 it was called "Plant Ecology."[11]

Wells's most interesting and personally rewarding extramural summer teaching was done in 1933 and 1934 at the New College Community. This communal educational institution occupied an old farm of eighteen hundred acres located twelve miles from Canton, North Carolina, on the east fork of the Pigeon River and at the foot of Mount Pisgah. It was operated by New College, a branch of Teachers College of Columbia University, which was a controversial experimental school for the preparation of teachers. It offered a non-traditional three- to five-year program leading to the bachelor's and master's degrees. Enrollment was strictly limited to a fixed number of students carefully screened for "good health, sound scholarship, desirable personal qualities, and promise of unusual growth." In addition to stressing close and informal relations among faculty and students in all aspects of daily life as well as in class situations, the program required periods of communal work on the North Carolina farm or in industry, as well as some months of foreign study and a year of teaching internship in a cooperating school. New College enjoyed only a brief life, for seven years from 1932. It was supported entirely by student tuition, but the enrollment never reached the projected numbers and the budget showed annual deficits. Citing financial reasons but not the radical political orientation of many of its participants, the dean of Teachers College abruptly terminated New College in 1939. Years later the head of New College claimed that the budget deficit had never been very large and that the dean had acted almost entirely for reasons of university politics.[12]

Some of the problems New College encountered were generated by its own personnel, all of whom were reacting against traditional aspects of formal education. The highly intelligent students of New College were largely from cities, and the faculty, selected for imagination and skill at innovation, were generally liberal in their thinking. Most of these people were different from the mass in ideas, talk, behavior, and dress; some were extreme. For that reason, the program received suspicious, sometimes hostile, attention in New York as well as in rural North Carolina, in professional educational circles as well as in the populace at large. There were rumors of communism, subversion, and immorality. In this regard, New College was probably largely the victim of the imaginations of its critics, although some of its individuals may have fitted the stereotype. Wells later recalled, for example, that one of the art instructors, who was an avowed communist, always managed to include a hammer and sickle somewhere in each of his paintings of plants and landscapes. On the other end of the political spectrum, however, a home economist participating in the program wrote glowingly of her experiences in Nazi Germany and declared, "After spending a few months in Germany one tempers one's remarks about Germany not paying her war debt and Hitler being a menace to the world." In any case, the people of New College were not ordinary.[13]

New College Community was established in North Carolina in the spring of 1933; its location was selected because of its climate, its abundant resources for the study of nature, and its remoteness from city life. Wells became associated with this communal enterprise through his wife, Edna. The Raleigh high school where she taught was engaged in a cooperative program with Teachers College by providing internship experience for student teachers. In participating in this program, she, and therefore also Wells, came to know Thomas Alexander, the inventor, organizer, and leader of New College. John Wilkinson Taylor, a close associate of Alexander's, also roomed in the Wells home during 1931 and 1932. Wells was instrumental in convincing Alexander to select the North Carolina site. His arguments must have impressed the Columbia authorities rather deeply, for an explanation of the choice attributed to an official of Teachers College is distinctly Wellsian in tone: "The mountains of North Carolina were favorably considered because of the remoteness from the hurry, bustle and strain of city life, the admirable character of the climate of southwestern North Carolina, and the rich biological, botanical, geological, geographical, and sociological resources of the area. The State provides not only the greatest variety of fauna and flora but also the best examples of each to be found within a like area in the United States." Similar language is found in other statements from New College. Wells also argued strongly for locating an additional program on the seacoast, but this suggestion was never taken up. Interestingly, in an article in the State College student newspaper which described these matters, the mountain farm operation was alluded to as "The Columbia University Mountain Summer School" rather than by its correct communal title. It may be supposed that Wells acceded to this designation, if he did not actually bring it about, because he himself later juxtaposed both names in his published description of the course he taught.[14]

In that account of his New College course Wells characterized it as an introductory treatment of botany with an ecological orientation: "From the very beginning the elementary student was made as equally conscious of habitat as he was of organism." This was accomplished by devoting the first three weeks of the three-month course to field studies on adaptation to environment in relation to the occurrence of plant communities in different places. It was a valid course at the college level in that the students studied in the laboratory the usual subjects of plant structure, function, and diversity; but one full day each week was devoted to a carryover of these subjects into field situations. These outdoor experiences included measuring habitat factors such as light and evaporation using suitable instruments. At the end of the course, plant communities and crops of the region were related to human life and industries of the area such as sawmills and paper mills. Wells worked hard in developing and executing this course; he cited his commitment to it as one reason why he could not devote any summer time in 1934 to a revision of *The Natural Gardens of North Carolina*. He was sufficiently enthusiastic about the outcomes of the course not only to publish a description of it but also to suggest that North Carolina colleges in general should consider instituting similar summer programs in botany, zoology, and geology for their students. His total experience in the New College Community impressed Wells deeply; over thirty years later, at the age of eighty-one, he wrote of it, "I look back on the two summers I had at the New College camp with the greatest pleasure. I visited Dr. Alexander and his wife there last summer." [15]

In addition to his description of the summer course at New College Community, Wells produced several other publications on subjects related to his teaching. They included two sets of laboratory directions for courses in elementary botany, one written while he was at the University of Arkansas and the other a collaborative effort with Shunk and Martin at North Carolina State College. The latter was used in temporary form in 1921–22 and in a printed version until about 1930. These laboratory outlines were sound but unremarkable. In later years the laboratory directions and other such course materials used at State College were mimeographed productions created collectively by the faculty, including Wells.[16]

Another publication, "A Method of Teaching the Evolution of the Land Plants," reveals some interesting aspects of Wells's approach to teaching. In this account he attacks "one of the *bêtes noires* of elementary botany instruction," helping a student to systematize details about different kinds of plants in such a way as to make clear the beautiful generalities of the evolution of plant generations. His method involves requiring the student to draw the life cycles of typical plants on a single diagram in concentric circles, the lowest in the evolutionary sequence near the center and so on out to the highest; equivalent structures of the various plants are placed on a single radius of the circles. The student is helped by the instructor with the first one or two cycles. "After that," Wells wrote, "he goes it alone. Acquiring his data from all possible sources he organizes it on his sheet where he cannot escape comparing the stages with those of the preceding types, with the delightful result that a goodly portion of learners really 'get the big idea' which is intended for them." Wells emphasized his belief that a student learns from this exercise because he does the organizational work himself: "For the instructor to make a large one (wall chart size) to be used as a basis for mastering the situation would be an unfortunate pedagogical error." To illustrate his discussion, Wells reproduced a diagram drawn by an actual student in one of his classes in the autumn of 1919, a figure that was "by no means perfect" but that showed "how far a youthful mind can go, provided it is given a logical start." Sixty-one years later, this student, Fred B. Monroe, vividly and proudly recalled his experience in making this diagram as a freshman and wrote to secure copies of it for himself, his brother, and his two sons. This fact strongly supports Wells's concluding claim that by use of his method "that veritable terror of alternation of generations has lost his Stygian aspect."[17]

In this paper on teaching evolution Wells confessed that "in fact the writer personally enjoys nothing more than directing working mentalities as they solve this problem for themselves." This statement, as much as any, reveals why Wells was a remarkable and effective teacher. He did not regard teaching as one of the distractions of academic life; on the contrary, he loved it. His influence was significant on a large number of students, some of whom recalled him twenty years after his retirement as "warm, inspiring, full of vigor, enthusiastic, dynamic, a character, quite a guy, like an older brother." He was outstanding in his formal classes, but he did not leave it at that: in many of his other activities, such as lecturing before laypersons and writing popular articles, he was functionally a teacher. Wherever he went, whatever he did, he was compelled to seize every opportunity to teach somebody something. Age and station were irrelevant; the prominent South African ecologist A. W. Bayer testified after meeting him in 1950, "Whether we were watching the dance of soil particles in a sand rift or discussing ecological principles, I was at the feet of a teacher who freely shared with me the thoughts of a more mature and abler mind." Wells himself said in 1955 that he regarded his teaching as his greatest contribution. Perhaps it was.[18]

CHAPTER II

Defender of Truth

The scientist has a conscience. He can no more violate his intellectual integrity by misinterpreting the simple open facts of nature to his fellowmen than he can lie to his fellowmen. . . . There is no valid excuse for the public denial of the facts of evolution. It has been nearly a century since evolution passed definitely from the realm of theory to that of established fact. And at the present time all scientists and impartial students of the subject are convinced that it is as much of an actuality as is the phenomenon of gravity. . . . Facts must be faced rather than dodged and in this connection we agree heartily with Bryan in his statement: "Truth is truth and must prevail!"—Statement of Professors, North Carolina State College, 1922

As an academician, Wells believed in complete freedom of thought: he believed that ideas should win or lose acceptance on their merits. So the first decade of his life in North Carolina was an intellectually turbulent one because it was just that for the state and nation. Science and technology, potent economic forces, and the aftermath of a tremendous world war combined with other factors in the 1920s to alter almost every thread of life. Consequently, large numbers of Americans feared the challenge posed by the fruits of change to their ancestral religious convictions. Most strongly they feared religious modernism, secular communism, and evolution, that supreme product of nineteenth-century biological and geological science. Evolution especially became to the fundamentalists, as they came at this time to be called, the crowning symbol of all they feared and hated, and they began to organize active opposition to it. Having learned from the government and from mass society during the hysteria accompanying World War I that intolerance and unreasoning hatred of foreign ideas might sometimes be justified, even desirable, the fundamentalists were determined to suppress the concept of evolution and to prevent the teaching of it by restriction, coercion, and legislation. They wished to propagate their own views by forcefully abridging the intellectual freedom of others.[1]

The antievolutionists generated controversy in many parts of the country, most famously with the notorious Scopes trial in Tennessee in July 1925. But they saw North Carolina as especially pivotal to their cause and so carried their fight to the state that regarded itself as "the Wisconsin of the South." Wherever he might have lived, Wells would have been their intellectual opponent; his personal views and makeup were such that it could not have been otherwise. His biology was too professional, his belief in democracy was too strong, and his conviction was too deep that every person should have the right and opportunity to shape his personal thought in the free marketplace of ideas. Because he was a professor in North Carolina and in the state capital, it was inevitable that he came to play an active role in the controversy. He was joined in this endeavor by Z. P. Metcalf, his close friend, neighbor, and professional colleague. Detailed accounts of the events of these years exist; however, Wells played more of a part in those events than most such accounts indicate.[2]

In the early 1920s the attacks of the antievolutionists centered on

William L. Poteat, president of Wake Forest College, who believed strongly in both Christianity and evolution and did not hesitate to express his views publicly. In addition to heading a Baptist college, Poteat was a professional biologist; at different times he was president of both the North Carolina Academy of Science and the Baptist State Convention. The assaults on Poteat were part of a general campaign in which evangelists from other states made extensive preaching tours around North Carolina thundering against evolution. Along with others, these men included such prominent personalities of the day as Mordecai F. Ham, Billy Sunday, and Baxter F. "Cyclone Mack" McLendon. Beginning in 1920, such evangelists staged annual Bible Conferences at the Baptist Tabernacle in Raleigh. Sponsored by the World's Christian Fundamentals Association, these week-long events were forums for the promulgation of antievolutionary views. At the third such conference in May 1922, Jasper C. Massee of Massachusetts preached against science and scientists in especially vitriolic terms. He contended repeatedly that scientists could not be Christians and that their principle of evolution was responsible for most of the ills of the time. He charged that the teaching of evolution would sooner or later destroy both Christianity and democracy, and he urged the people of North Carolina to withdraw support from all educational institutions that permitted the principle to be taught.[3]

This prolonged and vehement attack by Massee was too much for Wells and Metcalf, delivered as it was just a few blocks from their college. They drafted and delivered to a local newspaper a public reply signed by themselves and four junior colleagues, the botanists I. V. D. Shunk and A. C. Martin and the zoologists Clifford Otis Eddy and John Edward Eckert. Their published statement appeared with the heading "Reply to Massee about Evolution; Members of Faculty of N.C. State College Resent Remarks of Preacher." It was a long and carefully reasoned discussion of the scientific basis of evolution with examples of the evidence that compelled the scientific community in general to hold to the principle. In it they referred to themselves as Christian men who considered fundamental Christianity and science to be the two things of prime importance in the development of Western civilization. They stated: "To agree with Dr. Massee is the same as saying that a Christian has no business to study the evident and observable facts of nature." They further asserted: "Everyone

The still-vigorous B. W. Wells as he began his retirement, January 1955. Reprinted by permission of the *News and Observer* of Raleigh, N.C.

owes it to himself, to the cause of truth and to the intelligence with which he is endowed to give careful consideration to the simple facts of nature rather than turn one's face away from God's realities."[4]

At this point, in accordance with the previously arranged schedule of the Bible Conference, William B. Riley of Minnesota replaced Massee as the featured preacher. The day after the professors issued their statement, Riley responded to it with a public challenge to them to debate evolution with him. He declared that he was ready to meet any one of the professors "inside the confines of his own college or inside the Baptist Tabernacle" and proposed that the subject of the debate should be "Resolved, that the evolution synthesis is neither scientific nor spiritual." Wells and Metcalf found this opportunity for a public confrontation irresistible, and they forced the evangelist to proceed with it on their own terms. As indicated by his suggested topic, Riley wanted to assume the affirmative position; he also preferred to hold the event at the Baptist Tabernacle. But Wells and Metcalf were unyielding in their insistence that the debate take place at the college and that the subject be "Resolved, that evolution is a demonstrated fact." They demanded that the question be so put because they said that they were not opposing the Bible or Christianity or trying to undermine religious belief; they were simply insisting that the facts of biology and geology be accepted along with those of physics and chemistry. Because of the way he had put his challenge, Riley had to agree.[5]

The debate was set for four in the afternoon of Wednesday, 17 May, in the largest auditorium on the college campus; R. L. McMillan of the Pullen Memorial Baptist Church was named as moderator, with John A. Park and W. T. Bost designated as timekeepers. Wells and Metcalf together planned and drafted their talk, but they chose Metcalf as the spokesman. He was a leader in the First Presbyterian Church and was always cool under fire. Wells, by contrast, was not a regular church member; he also tended to become excited and to stammer when aroused, as he certainly was in this case. On the morning of the appointed day, a local newspaper ran a front-page story about the debate, complete with pictures of Riley and Metcalf. This account indicated that the event would be open to the general public and that the discussion would involve "no religious theories" and "strict adherence to the text of the question."[6]

The debate attracted a crowd estimated at two thousand persons; whatever its number, it overflowed the fifteen hundred seats in the auditorium of Pullen Hall. Thunderous applause greeted Riley's entrance, but it was more than matched by the partisans of Metcalf, mostly college students, when he arrived later. The debate lasted for an hour and a half. The agreed format allowed half an hour each for Metcalf and Riley in turn, ten minutes for rebuttal by Metcalf, fifteen minutes for Riley, and finally five minutes more for Metcalf.

As the debate developed, the approaches of the two men differed. Metcalf stood for evolution as science, while Riley ridiculed all that science offered. Metcalf's speech was described by the press as "coolly scientific," while Riley's was called that of "the trained debater."[7]

Metcalf opened with a carefully prepared speech; he spoke calmly and without aiming for oratorical effect. His introduction included the simple statements, "I am a Christian. I have accepted in its entirety the fact of evolution. I have never found anything in evolution to shake my fundamental Christian religion." He went on to define what scientists mean by evolution and then to summarize precisely the scientific evidence for it. He marshaled examples of changes in living organisms, the geological testimony inherent in fossils found in layers of rocks of different ages, the increasing complexity of animals and plants throughout time, and the features known as vestiges, which are the nonfunctional remains of formerly useful structures in organisms. In concluding, he challenged Riley to answer several questions directly and specifically.[8]

Riley's style in response was completely different. Speaking with facility and considerable animation but without a set text, he asserted flatly that the Bible was scientifically accurate, that Darwinism had been discredited, and that evolution simply had never occurred. According to the press, he shifted his attacks "with bewildering movement" from humorous anecdotes to "cryptic indictment." At one point he referred to some pictures of prehistoric men in a book on evolution: "Come up here after the debate and look at these pictures, and I am sure you will see somebody who looks like them when you get downtown. I am glad that the weak don't die. Some of you folks may develop into something yet if you stick around here for 500,000,000 years."[9]

In categorical statements delivered with complete assurance and conviction, Riley answered the specific questions Metcalf had posed. The difference in the approach of the two men is illustrated by three of those questions and answers. Metcalf: "Why do living organisms present themselves in such marvelous graded series, protozoa to man, bacterium to dandelion?" Riley: "That is the order of God's creation. He began with grass and ended with man." Metcalf: "Why do the higher organisms develop the nonuseful structures known as vestiges?" Riley: "Who said they were nonuseful? God may have a function for them that you have not found out. I have still got my appendix and I'm going to keep it. Those other 186 I never did have." Metcalf: "Why among the vast array of simple animals and plants known to have lived in the coal age not one flowering plant nor one mammal has been found?" Riley: "That is down where God began."[10]

In his rejoinder, Metcalf explained that Darwin's theory had to do with a possible mechanism of evolution, not with the fact of it. He also emphasized that evolution concerns the development of life, not its creation, which is a completely separate matter. And he pointed to Riley's apparent lack of scientific knowledge: "No scientist would declare that there is no life without blood. Go look for blood in yonder green maple tree." Riley responded in the vein in which he had begun and even returned to the maple tree: "Go in the spring and cut a ring around it, and see it bleed to death." By this time he was so enthusiastic that "it was a full minute before he heard the pounding of the time keepers." In concluding the debate, Metcalf stressed that although there were a number of interpretations of Darwin's theory of how evolution occurred, there were also many creeds that interpreted the Bible differently.[11]

At the conclusion of the debate, the crowd surged out satisfied. The news reporter who witnessed the event judged that it had been an opportunity to release the steam generated by a week of intense discussion in all quarters. "It was quite without parallel in the annals of polemics hereabouts," he wrote. The supporters of each speaker felt comfortable with the outcome although they all apparently came in and went out with the same opinions. L. A. Whitford, who was a student at State College at the time, wrote many years later, "I heard the debate from the south balcony and thought science won, of course." The Metcalf-Riley debate received statewide and even national attention at the time and has since been duly noted by students of the evolution controversy of the 1920s. But most writers on the subject have failed to recognize the important part played by Wells, probably because as the actual participant Metcalf was more visible to the public.[12]

After 1922 antievolution pressure continued to grow in North Carolina, mainly through attempts to prevent the teaching of the principle. There was a concerted effort, which failed, to require the teaching of Bible courses at publicly supported colleges. In 1924 Governor Cameron Morrison forced the state Board of Education to strike two biology textbooks from the list of approved schoolbooks on the ground that they contained discussions of evolution. "I don't want my daughter or anybody's daughter," the governor said, "to have to study from a book that prints pictures of a monkey and a man on the same page." In 1925 strong public outcries arose when Albert S. Keister discussed evolution in a class at the North Carolina College for Women, later the University of North Carolina at Greensboro. And in the same year considerable objection was voiced to the choice of W. L. Poteat to deliver a series of special lectures at the university at Chapel Hill.[13]

During this time Wells also was not idle. He instigated the passage of a resolution on evolution by the North Carolina Academy of Science at its meeting in May 1924. On the first day of that meeting he got himself appointed chair of the resolutions committee; the other members were Collier Cobb, a geologist at the University of North Carolina, who was an open defender of evolution, and Lula G. Winston, a professor of chemistry at Meredith College. On the second day of the meeting, Wells presented a draft resolution on evolution which evoked heated debate among the assembled scientists. This proposal consisted of a preamble and three sections: the first stated that biologists regard evolution as an established fact, the second asserted that controversies among scientists concerning Darwinism had not discredited evolution itself, and the third declared evolution to be "merely the regular and ordinary method by which God works in the natural world." Discussion of the resolution did not proceed for long before an attempt was made to adjourn the meeting without taking any action. But the chair entertained and gave precedence to a motion to adjourn and reconvene in two minutes; this motion carried. Amid considerable hubbub some opposing strategies were quickly formulated. When the meeting resumed, a motion was introduced to refer the resolution to a committee that would study it and report the following year; this measure was defeated. The preamble and first section of the resolution were then approved with some changes in wording, but the second and third paragraphs failed to carry. The resolution as adopted thus read as follows: "Whereas, among the many attacks that have recently been made on evolution, there frequently appears the statement that 'evolution is a discarded and discredited theory' and whereas this statement is totally incorrect, therefore, the North Carolina Academy of Science hereby declares that to the best of its information and belief, practically all scientists look upon evolution not as theory, but as an established law of nature." This resolution still stands as an official position of the academy.[14]

Although it was not obvious at the time, the climax of the evolution controversy in North Carolina was reached in the state legislature in 1925. On 8 January Representative David Scott Poole introduced Resolution Number Ten into the House of Representatives. The text of this resolution, which was referred to the education committee, was as follows: "Resolved by the House of Representatives, the Senate concurring, that it is the sense of the General Assembly of the State of North Carolina that it is injurious to the welfare of the people of the State of North Carolina for any official or teacher in the State, paid wholly or in part by taxation, to teach or permit to be taught, as a fact either Darwinism or any evolutionary hypothesis that links man in blood relationship with any lower form of life." Legislators opposed to this measure asked President Brooks of North Carolina State College to coordinate the efforts of state institutions against it, but he declined the honor and conveniently absented himself from Raleigh at the height of the fight. President Poteat of Wake Forest College worked behind the scenes but avoided public involvement because the resolution concerned only tax-supported institutions. President Harry W. Chase of the University of North Carolina assumed public leadership and personally bore most of the attacks of the antievolutionists. But Wells and Metcalf worked closely with Chase in planning the opposition.[15]

The committee hearing on the Poole Resolution was held on 10 February before a crowd so great the session had to be transferred to a larger chamber. The antievolutionists spoke first, then President Chase delivered a lengthy and eloquent speech in opposition to the resolution and in defense of academic freedom of teaching. Metcalf testified next, followed by Wells, "both of whom put themselves down squarely as Christians, members of the church and believers in evolution as fact." In his statement Wells first read the resolution of the Academy of Science and then said, "I stand here to deny that a man cannot believe in evolution and be a Christian. I am here as a living exponent of the Christian religion in all its essentials and also in evolution as fact." After pointing out that the technological fruits of science are evidence that its facts are not imaginary, "he denied that in teaching botany he ever did more than present the facts as they have been found to his students and said he left the matter then with the student. He was a Christian, a member of the church, and his belief in evolution did not make himself less of a Christian but more of one." In response to a question as to when the soul enters the life of a man if he started millions of years ago, Wells "declared it made no difference to him when that happened." The testimony of Wells and Metcalf received favorable comment in the student newspaper at State College: "Their speeches were very effective and bore much weight judging from the applause of the audience."[16]

After hearing arguments and deliberating, the committee voted eighteen to seventeen to report the resolution unfavorably, the chair casting the tiebreaking vote. Because of the closeness of the decision, a minority report was filed to ensure a vote in the full House. During the week between the committee action and the floor debate, President Chase worked hard to organize opposition through speeches, conferences, and correspondence. Wells was involved in this activity; he received one letter in which Chase stated that "we should be ready to turn loose on the Senate" if the resolution passed the House. In fact, after lengthy debate, the Poole Resolution was finally defeated in the House on 19 February by a vote of sixty-seven to forty-six. Later Wells and Metcalf notified the scientific community of these legislative events in a letter to the journal *Science*; after quoting the text of the resolution, they simply summarized the facts attendant on its defeat.[17]

It was not clear at the time, but the defeat of the Poole Resolution in the legislature marked the beginning of the end of this particular evolution controversy in North Carolina. The antievolutionists continued to aim at public schools and also set their sights on the next legislature, which would meet in 1927. They managed to induce school boards in several counties to ban textbooks that described evolution. In May 1926 they organized the Committee of One Hundred to "make our schools safe for our children"; this organization was later known as the North Carolina Bible League. In the same month outside evangelists came into the state and set up the Anti-Evolution League of North Carolina. Other activities included an ineffectual debate in Charlotte; also, antievolution resolutions were passed at the meeting in North Carolina of the education board of the Southern Baptist Convention and at the convention of the North Carolina State Federation of Labor. A strong but unsuccessful attempt was made in the North Carolina Baptist State Convention to bar the teaching of evolution at Wake Forest College and to remove President Poteat. The antievolutionists had been greatly angered by the Metcalf-Riley debate and by the outspoken testimony of Wells and Metcalf before the legislative committee. They therefore exerted great pressure on President Brooks, demanding that he purge North Carolina State College of "such obvious heretics." To his credit, Brooks refused their demand.[18]

Wells also continued to be active in the controversy. At the meeting of the North Carolina Academy of Science in April and May of 1926 he proposed from the floor another resolution which would place that organization strongly in support of freedom of teaching and therefore of evolution. This measure was referred to a committee that included Wells, William Chambers Coker, a botanist at the University of North Carolina, and Bert Cunningham, a zoologist at Duke University. The committee made the resolution even stronger by adding a provision in support of Chase and Poteat: "The North Carolina Academy of Science desires to reiterate that if the present rate of progress and enlightenment in the State of North Carolina is to be maintained and advanced, it is absolutely and unqualifiedly necessary that all those hypotheses, theories, laws and facts which constitute the legitimate content of any field of study, may be dealt with at any time by teachers. The Academy goes on record as endorsing most emphatically the stand of Dr. H. W. Chase and Dr. W. L. Poteat on the freedom of thought and teaching." A standing vote was called for, and the resolution was adopted unanimously.[19]

In July of the same year, Wells again wrote to the scientific community through the journal *Science*. In this letter he described in detail the aims of the Committee of One Hundred (North Carolina Bible League). His general assessment was that the situation would be much more serious in the coming legislature than it had been in 1925; he felt that the state institutions of higher learning faced a "theological raid on education" and a "modern Protestant holy inquisition." Because Duke University and Wake Forest College were standing firm for freedom to teach evolution, he ironically pointed out that "in case the fundamentalists swamped the state institutions, it would become desirable, to prevent 'the ruin of youth,' to transfer the young men in the church schools to the 'safe' state fundamentalist colleges." Wells also wrote to the president of the Science League of America, Maynard Shipley. In a letter of 11 December 1926 he advised Shipley that the antievolutionists were "already again astir in this state and getting their forces organized for a new attack on the coming legislature which meets in January."[20]

But this time Wells overestimated the strength of the foe, for the antievolutionists had largely failed in the elections of 1926 and their efforts in the legislature were foredoomed. In January 1927 Representative Poole did introduce into the House a bill more stringent than his resolution of 1925. This measure would have banned the teaching in any publicly supported school not only of evolution but also of any doctrine or theory contradicting the biblical account of human origin. Violation of this prohibition would have been punishable by fine or imprisonment and by disqualification from teaching. In February, however, the committee on education defeated the bill by a vote of twenty-five to eleven; a minority report was filed, but the measure was not called up on the floor of the House because its defeat was certain. Neither Wells nor anyone else from a state institution of higher learning appeared to testify at the legislative hearing on the bill. In so refraining, Wells was complying with the wishes of his college president. At a meeting of the Faculty Council on 1 February Brooks "urged that unless it became absolutely necessary, that State College should not be drawn into the discussion of the bill." In fact, such involvement was not necessary, for with the vote of the committee on education the active antievolution movement was dead in North Carolina, at least for the time.[21]

Throughout the decade of the 1920s Wells contributed significantly to the struggle against anti-intellectualism by correspondence, in behind-the-scenes planning, through the Academy of Science, and by plain public statement. It is striking that while those who dared to speak from Wake Forest College and the University of North Carolina were the presidents of those institutions, the public voices from State College were those of Wells and Metcalf. Wells was indeed a vigorous proponent and defender of scientific truth and freedom of thought. His makeup was such that he could not do otherwise; by profession he was an academician, and he was always professional about it. He brought to his students, his department, his college, and his state a total commitment to the life of the mind. In this regard he influenced all of them in ways that cannot be measured but may still be felt. In addition, he was not afraid to step outside the academic world and expose himself publicly to harsh criticism if he felt that the intellectual way required defense. He certainly could not always have been right. But his ideals were always right: science, scholarship, excellence, the university, truth.

PART 4

Private Person

CHAPTER 12

Early Years, 1884–1919

To the young high school freshman walking along the creek bank, the brilliant red flower presented a challenge. He didn't know what it was, but he had been taking botany for a whole week, and he had under his arm Wood's *Manual of Plants*, a volume which could be used to identify flowers. And he felt he had learned how to use it. He bent over the flower, studied it, and began using the key in the manual. Before long, he located a description which fit perfectly and found that he was looking at *Silene Virginica*, known popularly as Fire Pink. "From that moment on," says Dr. B. W. Wells now, "I knew what my life's work would be."

—Herbert O'Keef, 1955

SPANNING MORE than nine decades, the personal and private life of Wells was as full and as eventful as was his professional and public one. That life began on 5 March 1884 in Troy, Ohio, when he was born the youngest son in a family of three sons and two daughters. His father, Edward T. Wells, was a Methodist clergyman, and his mother, Lucia Morehouse Wells, was even more devoutly pious than her husband. The Wells home was closely governed by the religion of the parents, which pervaded all family activities. The most acceptable and most frequent topic of conversation at mealtimes was the religion of the Holy Bible as the parents interpreted it. Wells experienced his childhood in a stern and authoritarian atmosphere. His parents cared deeply for their children but insisted on strict regulation and made little display of affection. The influences of this early home life persisted in Wells through all his years. He learned well and accepted the lessons of honor, responsibility, duty, and morality, but he rebelled inwardly at an early age against stifling restriction. He was particularly repelled by the kind of religion preached by his father and subscribed to by his mother; he later declared that in reaction he became a religious agnostic at the age of sixteen.[1]

But other seeds were sown in that early home which brought Wells pleasure throughout his life. His mother possessed artistic and musical abilities, which he inherited. An interest in art was awakened, which led to the painting activities that contributed such a satisfying element to his mature years. In music he never had formal instruction, but he took naturally to playing the piano by ear and years later acquired a flute, which he learned to play passably. His sense of rhythm was excellent, and in mature life he became a superb dancer who loved to engage in that activity.[2]

Wells's youth was spent in Wilmington, Lockland, and Dayton, Ohio, as his father moved from church to church in the Methodist manner. He attended public schools. At Steele High School in Dayton he studied botany as a freshman, an activity that determined the principal course of his life. One day, a week into that course, he spied a brilliant scarlet flower while walking along a creek bank. Strangely struck by it, he opened his manual of plants and attempted his first identification by means of a descriptive key. A thrill of triumph surged through him when he successfully found that the plant was known as the fire pink. Never mind that the plant was a common one, never mind the technicalities of oblanceolate cauline leaves, pilose tubular calyx, deeply cleft petals, and the rest. He had found its name by himself, wresting it from the book full of long, hard words, using his eyes to see and his head to think. What had been just a flower, a beautiful but strange flower, was now a part of him because he knew its name and would recognize it again anywhere. Excitement almost overwhelmed him, for he suddenly saw the future, as if a huge door had burst open: standing in that Ohio wood he knew what his vocation would be. He would study plants, he would be a botanist. He never forgot that plant or that instant of flashing awareness. Years later he included a picture of the flower in *The Natural Gardens of North Carolina* with the personally significant caption, "The brilliant red of the fire-pink always attracts attention." And at the age of seventy he recalled, "From that moment on I knew what my life's work would be. I was so terribly excited about identifying that flower. And I'm still as excited as ever about botany. The more you see, the more you have to see."[3]

Another incident from Wells's youth always remained vivid in his memory. At some point he joined the Ohio National Guard, and when President William McKinley was assassinated in 1901, his unit, along with many others, was called to Canton, Ohio, to serve during the period of the funeral and entombment. In addition to marching in the funeral procession on 19 September, the six thousand guardsmen positioned themselves along streets and around the McKinley house, where the president's remains lay during the night before. According to contemporary accounts, the crowds that massed in the vicinity of the house were enormous, amounting to tens of thousands. "North Market Street was a living, seething mass of humanity for five squares below the house and for three squares above. Several regiments of soldiers were required to preserve a semblance of order." The seventeen-year-old Wells was one of those who stood around the house in a triple line from the curbs to the lawns. Understandably, he never forgot the experience of facing such a crowd with bayonet fixed.[4]

When Wells completed his public school career by graduating from Steele High School, he enrolled in the Ohio State University at Columbus to fulfill his aim of studying botany. He was a hard worker and an excellent student whose scholastic attainments were eventually recognized by membership in the honorary organization Phi Beta Kappa. Although his primary emphasis was on his studies,

Teenaged B. W. Wells at home with his stern and devout preacher-father, Edward T. Wells, about 1900. Courtesy of Maude Barnes Wells.

he also participated in extracurricular activities; one of these involved marching with the flag at football games. He provided for his college education with his own earnings, a fact of which he was always proud. The only money he received otherwise was a sum of seventy-five dollars advanced by his father; he repaid that shortly after his graduation.[5]

Wells had hardly begun his college studies when he was forced to interrupt them. During his first year, he had to drop out to help support his family. This step became necessary when his father lost the use of his voice and could not continue his pastoral duties, especially preaching. After only a few weeks, the congregation, an affluent one, dismissed the father from his pastorate and turned the family out of the parsonage, leaving them in a precarious financial condition. His father later recovered but never took another church, instead making a comfortable living as a real estate broker. Wells was away from the university for four years. During this time, he held menial jobs: he worked in a stove factory, where he once was struck by heat prostration; on an orphanage farm; and as a waiter in a hotel, where he once experienced the embarrassment of inadvertently spilling hot soup down the back of a prominent Republican politician. He ended this phase of his life with two years in Toledo as a highly successful agent for a large property company; using a motorcycle for transportation, he collected tenement rents in a district of ill repute. But he never changed his goal or shelved his academic plans more than temporarily. During this period of marking time, he gathered and studied plants on his own, an activity for which the botanists at the university later awarded him some college credit.[6]

Wells's early home and family life had led him to reject the outward form and much of the dogma of the parental religion. The harsh treatment of his family by his father's last congregation thoroughly fixed this rejection; for the rest of his life he had no interest in and little respect for the organized Christian church, of which he was never subsequently a member. These early experiences produced in him an unusual mixture of convictions. Despite his distaste for the church, the ethical code in which he believed and by which he lived was as strong and as fundamentally Protestant as if he were the most pious of devoted church members. He was not altogether opposed to religious organization: later in the 1920s he was a moving spirit of a group that organized in Raleigh a short-lived nondenominational church which was extremely liberal by any standard. This organization evoked much critical comment in the community; indeed, the views contained in one of his public talks before this group generated considerable outside protest and even open suggestions that he should resign his professorship. During the evolution controversy of the same period, Wells could honestly testify before the committee of the state legislature that he was a church member and a Christian, "a living exponent of the Christian religion in all its essentials." Given that he himself was deciding what those essentials were, his statement was true. It was misleading, perhaps, but he was perfectly sincere in delivering it, for he believed in the Christian ethic if not in the form and ceremony of the church. And his personal actions were as consistent with his beliefs as he could make them. In later years a professorial colleague categorized Wells as "the most Christian atheist" he had ever known.[7]

In the fall of 1908, Wells finally returned to Ohio State to resume his college program. He continued there for three years and received the A.B. degree in 1911. These years in Columbus were rich ones for his education, especially in botany. During this undergraduate period, he experienced the special satisfaction of seeing his first scientific publication appear, a study of the development of certain marine algae.

He also quickly came under the influence of John Henry Schaffner, an outstanding professor who achieved international renown. Wells considered this man to be one of the greatest botanists of his time and thirty years later still felt that to have studied with Schaffner "was one of the special privileges" of his life. He vividly recalled working with him in the decrepit and disintegrating botany building, which leaked water with every rain. Despite the deficiencies of the building, the professor carried on his botanical study with an enthusiasm which, Wells testified, he conveyed to his students, "who sometimes needed it when they too suffered from the rain or cold." Wells thought that his own biological insights were greatly deepened and his horizons greatly broadened through many scientific and philosophical discussions with Schaffner. It is possible that the professor helped the young Wells come to grips with his mixed feelings about religion, for R. F. Griggs reported seeing Schaffner "sit by the hour discussing philosophy and religion with some student who after being brought up in a narrowly orthodox home found himself all at sea in the free atmosphere of the university."[8]

As an undergraduate student Wells participated actively in the Biological Club, which included in its membership professors as well as students at all levels. This club was a serious and professionally oriented one, meeting regularly for discussions of research pursued in Columbus and elsewhere; the minutes of its meetings were published at intervals in the *Ohio Naturalist*. Wells was elected a member of the club in November 1908 and was active in it from then on. In 1910 he spent part of the summer at the Lake Laboratory, a facility operated by the university at Cedar Point near Sandusky, and that autumn reported on his experiences to the club. In November of that year he was elected secretary-treasurer of the organization and thereafter regularly composed its minutes. During this academic year, the club engaged in a study of the history of biology, with various members undertaking to examine and report on different periods. On 13 February 1911 Wells presented the first of this series of papers by discussing the early history down to the time of Galen in the second century.[9]

During his undergraduate years, Wells formed an unusually close association with three other students also destined to make their marks as professional biologists. Bentley Ball Fulton became an entomologist and from 1928 to 1954 was also a professor at North Carolina State College; Joseph Lyonel King was later a leading entomologist with the United States Department of Agriculture; and Wencel Jerome Kostir was a longtime professor of zoology at the Ohio State University. These four young men roomed in the same house during Wells's senior year; they were kindred spirits and lifelong friends, even though all four met together only three times after 1911, the last time in 1944. Robert Fiske Griggs, a noted ecologist, who was then assistant professor of botany at Ohio State, regarded the four as among the best students at the university and introduced them to the natural history delights of the Sugar Grove district, a remote and then wild section of Hocking County, Ohio. Wells himself reported to the Biological Club on a trip he took there with Griggs during the spring vacation in 1910. The four friends visited the area frequently on their own, sharpening their biological skills. In the summer of 1911 they rented an old log cabin and a hay barn at the head of Conkle's Hollow and set up what they called "The Big Pine Biological Laboratory." For ten weeks they lived under primitive conditions, tramping the woods and studying nature according to their own interests. This unique experience not only cemented their friendship but also formed an extremely valuable part of their biological educations.[10]

Upon graduation in 1911, Wells held only a bachelor's degree, but at age twenty-seven he was older than most fresh college graduates; he was also more experienced in the world and in science. He was thus able to obtain teaching positions, albeit junior ones. For the academic year 1911–12 he was instructor in biology at Knox College in Galesburg, Illinois, a small, nonsectarian, coeducational school. The following year, 1912–13, he served temporarily as head of the Department of Botany at Connecticut Agricultural College, now the University of Connecticut, in Storrs. In this position he substituted for the prominent geneticist Albert Francis Blakeslee, who was on leave for a year at the Carnegie Institution at Cold Spring Harbor, New York, a noted biological research establishment. In Storrs he at first rented a room in the absent Blakeslee's house. When he learned that the rent charged him was greater than that paid for the entire house, in a typical Wellsian response he immediately moved to other lodgings in neighboring Eagleville, even though that action necessitated long walks to work, sometimes through deep snow. After leaving Connecticut he served for two years, 1913–15, as assistant professor of botany at Kansas State Agricultural College, now Kansas State University, in Manhattan.[11]

Wells's experience in Kansas affected his later life in two important ways. His abilities so impressed authorities at Kansas State that they later strongly recommended him to North Carolina State College. It was largely through this influence that he secured the position in which his major life's work was done. Also, in one of his classes he encountered Edna Metz, a young woman from Jewell City, Kansas. Metz was a student of science, and she so struck Wells that he mounted a deliberate campaign to woo her. His effort succeeded, and later the two were married.[12]

For complete realization, Wells's dream of becoming a botanist required that he secure advanced degrees. Therefore, in 1915 he returned to the Ohio State University to embark on graduate study. He had by this time become fascinated by the insect galls of plants; he had actively collected specimens of these in Ohio and while teaching in Connecticut and Kansas and had already published two accounts of his studies. He now conducted his thesis research on galls of the hackberry tree under the guidance of Professor Schaffner; he received the M.A. degree in 1916. His professional ties with his home state continued for many years thereafter. He had become a member of the Ohio State Academy of Science in 1911, a relationship that he maintained at least until 1930. And in 1917 he was awarded a grant from the research fund of that academy.[13]

To continue work for the highest graduate degree, Wells moved in 1916 to the University of Chicago, which at that time had one of the leading botanical faculties in the country. He was formally directed at Chicago by the ecologist Henry Chandler Cowles; in actuality he worked more or less independently, for neither Cowles nor any other professor there knew much about insect galls. He completed his advanced study and research in one year and received the Ph.D. degree in 1917. In August of that year he married Edna Metz. By the time he completed his formal education and embarked on matrimony, he had already reached the relatively advanced age of thirty-three.[14]

Wells was armed with a new doctorate and a new wife, but the academic position he needed to fulfill his dream of a career in botany was slow in coming. The nation and its institutions of higher learning were in the midst of upset caused by World War I, and the best post he could secure was as professor of biology at Grubbs Vocational College in Arlington, Texas. That school was a technical junior college administered through the Texas Agricultural and Mechanical College. Grubbs College was later named the North Texas Agricultural College; in 1959 it was made a four-year school and in 1965 became the University of Texas-Arlington. Wells's stint there was painful to him. The school was in its first year of operation, and he was horrified by its poor facilities and shoddy academic standards. His reaction to this experience reveals an interesting aspect of his personality. His pride in his own worth was considerable, and he felt humiliated to be associated with any enterprise he thought was of low quality. Hence for a great many years he kept secret his year at Grubbs. In all of his published biographical sketches, such as those in *American Men of Science*, that year of his life was simply a blank, as it was in all of the summaries he kept privately. But very near the end of his life, certainly at least as late as 1977, he apparently concluded that Grubbs had achieved reputable status. On a typewritten biographical summary found among his papers there appeared this penciled addition in his handwriting: "Professor of Biology Texas State Univ, Arlington, Tex. First year 1917–1918."[15]

When the opportunity arose, Wells gladly moved to the University of Arkansas in Fayetteville, where for the academic year 1918–19 he was professor of botany and head of the department. This was a step up, but not enough of one to suit his ambition. The institution was poorly supported by the state; a survey conducted at about this time by the United States Department of Education reported that although the faculty was "well selected," "the poverty of the University is most apparent in its plant and equipment." In addition to this basic deficiency, the atmosphere during Wells's year there was poisoned by other factors. One of these was unrest occasioned by a lingering dispute between the president of the institution and certain factions of the student body and the public. Another was the war, which disrupted academic life considerably: one historian reported that "in the fall of 1918, under the Student Army Training Corps regime, much of the campus became an army post patrolled by armed sentries and controlled by regulations which made usual college work extremely difficult and at times impossible." In addition to botany, Wells had to teach courses in hygiene and sanitation for military students. Finally, the epidemic of influenza which swept the country that year caused the deaths of fourteen students and delayed the beginning of class work for more than a month. All in all, Wells was not happy at Arkansas and was delighted to accept the post at North Carolina State College when it was offered to him in 1919. Thus his academic wanderings ceased, and he joined the institution where he worked for so many years and on which he left such a mark.[16]

CHAPTER 13

Raleigh Years, 1919–1954

One of the earliest things was magic tricks. He was babysitting me . . . and there was a beautiful little pebble in the driveway, and he picked it up and said, "You need a piece of candy, don't you, Katherine?" and I said, "Yes, *sir*!" And with just a quick motion I had a piece of candy. So then I went and got everybody else in the neighborhood to come and see Dr. Wells make stones into candy. . . . All of my life Dr. Wells and daddy saw to it that I was included. . . . I remember working part of the time for Dr. Wells, collecting plants and putting them just right, and part of the time for my daddy, opening the bottles and putting the bugs in so they were just right, and I thought I was so important. . . . He was an artist in every way, shape, and form, and a scholar, and a second father, and, I guess, one of the greatest friends anybody could have.—Katherine Metcalf Browne, 1978

WELLS FOUND his thirty-five years of life in Raleigh fulfilling in many ways. He had, of course, his work as professor, researcher, and administrator at North Carolina State College; these would have been enough to occupy any ordinary man. In addition, he pursued his interests in conservation and popularization of nature, activities which involved a steady schedule of speaking engagements, writing, and other commitments. But his mind was continuously active and required almost constant stimulation. As a consequence, he filled every available corner of his personal life with enterprises of various sorts and participated fully in the life of the world around him.

Wells possessed tremendous intellectual enthusiasm. He was very intense, a dynamo of thought, a mental live wire from which sparks flew. The range of his interests was broad, including not only sciences in addition to botany but also literature, art, music, history, politics, and contemporary social conditions. Because of a curiosity about almost everything, he was an insatiable reader, eager to discuss all that he read. Through most of his life his students, friends, and colleagues were amazed at the breadth of his knowledge; rarely could a subject be introduced on which he could not comment intelligently, and rarely did he refrain from so doing. Merely to associate with him was tremendously instructive. Indeed, many years later a biologist from Wake Forest University testified that at scientific meetings conversational hours spent with him and others like him were more educational than the formal programs.[1]

Wells loved to talk, and he would talk freely with almost anyone he met. He relished thought, and he thrilled to the enjoyment of meshing intellects with others in the discussion of important ideas. For more than thirty years he was active in the Sandwich Club, a discussion group composed of prominent Raleigh figures. Open by invitation only, its membership included, among others, businessmen, physicians, an Episcopal clergyman, a well-known journalist, the president of a women's college, and an eventual governor of the state. These men of interestingly mixed backgrounds shunned publicity concerning their group; in fact, they tried to keep its existence relatively secret. Their sole purpose was to meet at regular intervals for their own personal pleasure and edification. At first, they enjoyed each time a simple meal, from which the name of the group was derived, but later such repasts became elaborate dinners. The discussions in which this group engaged thoroughly delighted Wells, and he was very proud to be a member.[2]

Wells's intellectual dynamism was matched by his physical vitality. Although not a large man, he was wiry and so full of physical vigor that he seemed large to others. One woman who encountered him only once pictured him at the age of eighty-four as "a big and spirited man." In pursuing his ecological fieldwork he faced the outdoors squarely, confident in his own strength. For two summers on the Big Savannah he and his co-workers lived in a rented old shack, which they used as a camp and field laboratory and dubbed "The Crystal Palace." Here a kerosene lantern provided illumination and a kerosene stove the means of cooking. Water came from a hand pump in the yard, and baths were taken on an old barn door. This spartan life was eased somewhat by frequent evening meals at a boardinghouse in nearby Burgaw. During his study of the vegetation of the sandhills, field accommodations were even more primitive, consisting of cots in an old army tent, a stove, and waterproof boxes for storage. In his later work on the grassy balds at the age of fifty, he spent two summers hiking mountain trails and climbing peaks, usually alone and frequently at distances from his car of more than ten miles. Even into his eighties he regularly led or participated in field excursions which most persons found somewhat grueling. As an old man, he periodically went up Mount Mitchell with groups of graduate students, and he was surprised to note that as the years passed fewer and fewer of them could match his pace on the twelve-mile up-and-down trip. He carried this physical vigor into his personal life as well, for he was a superb dancer and loved to engage in that activity. Evenings of ballroom dancing at the Wells home were occasional occurrences in the 1930s, and during his two summers at the New College Community he acquired a love of square dancing. The fall field trips with his plant ecology classes usually included a Saturday night stop at Carolina Beach, and his ability to outlast anyone in the street dances there contributed to the awe with which his students regarded him. Indeed, throughout his long life Wells's energy impressed many persons of all ages, whom he worked, walked, or danced almost to exhaustion.[3]

Thought, discussion, and outdoor and indoor physical activity were not Wells's only pursuits in his personal life. He was always interested in art, although he never found time to pursue it inten-

B. W. Wells and his second wife, Maude, in the yard of their home on Park Drive in Raleigh, October 1944. Courtesy of Maude Barnes Wells.

sively until he retired professionally. As a youth he was skilled in drawing with pencil or pen and ink. When he was out of college as an undergraduate, working to help his family, in addition to studying plants he took a correspondence course in drawing. Back in school, he and his college friend Bentley Fulton together used the preparation of laboratory notebooks as a means of sharpening their considerable skills in drawing. Subsequently the line engravings accompanying some of his publications on insect galls evinced a drafting ability rarely seen in scientific papers today. He often said that he had been the unacknowledged draftsman of a well-known and frequently reprinted diagram illustrating plant succession in bogs; this diagram appeared in a publication by Alfred Paul Dachnowski, whom Wells met while a student at Ohio State. In his mature years he tinted by hand many of the glass lantern slides used in his public lectures, and by the summer of 1932 he had progressed with oil painting to the point of enrolling in two art schools in New England. He also designed the cover for the original edition of *The Natural Gardens of North Carolina* and made the drawing of the Venus flytrap that was reproduced thereon. He eventually became confident and proud enough of his own abilities to list himself in a personally prepared biographical summary as a "gifted amateur artist."[4]

Wells did not keep his interest in art solely to himself. He gave public talks on art and even incorporated that subject into some of his lectures on plants. One such discussion given in 1936 had the title "Some Principles of Art in Relation to Garden Planting." In the course of this lecture, "using a large easel covered with newsprint he sketched out his plan as he talked, developing each separate phase of planting into the finished pattern." During the 1940s, he interviewed various artists on radio programs broadcast by the local station WPTF. Finally, he was a longtime member of the North Carolina Art Society, perhaps from its beginning, and once served on the board of directors of that organization.[5]

Wells's interest in and feeling for art also found its way into his academic work at North Carolina State College. In 1934 the report of the college curriculum committee, of which he was general chair, emphasized the importance of art in general education: "It is becoming increasingly evident that one of the weaknesses in higher education, and especially characteristic of technical schools, is the lack of adequate opportunity for the students to develop in art appreciation. The artistic sense and judgment of most college graduates is so low that pictures and decorations amounting to atrocities are placed in their homes as objects and designs of which they are proud. The aesthetic natures of these college selected young people are almost wholly neglected." As part of a more general emphasis on the importance of cultural values in a technical curriculum, the committee recommended offering as a start a few courses in art and art appreciation open to all students.[6]

Wells's interest in art, including modern or nouveau art, also involved him in the affair of the McLean murals, a controversy that alternately blazed and smoldered at North Carolina State College for almost seven years and reached a resolution only half a century later, in 1982. The library committee at the college in 1934 commissioned Raleigh artist James Augustus McLean to paint four large panels to be hung in the rotunda of the library building (later known as Brooks Hall). Begun in May and completed in November, this work was supported by the depression-era Federal Public Works of Art Program, with materials provided by the State College Women's Club. Working according to designs previously approved by the library committee, McLean produced four separate panels representing agriculture, architecture, engineering, and science. In planning these murals, especially the one depicting science, McLean discussed their content with various individuals, including Wells, whose own artistic efforts he had previously aided; Wells's suggestion, for example, that the atomic nucleus was a source of energy along with the sun was heeded by the artist, who incorporated a crystal of uranium in his design.[7]

McLean's paintings were first displayed publicly at a meeting of the North Carolina Art Society in December 1934. In connection with this exhibition, Wells published a long and detailed newspaper article complete with pictures of the four panels and of the artist. After describing the contents of each of the paintings, he discussed at length their artistic qualities. He approved of them in enthusiastic terms, summarizing his appraisal as follows: "The McLean murals stand high in their field of art. In symbolical content, they show painstaking research and resultant accuracy of interpretation. As artistic expressions they are characterized by marked individuality of treatment, a refined judgment in the choice of his pigments, high success in composition, and a masterly use of the thin color so desirable in mural work. North Carolina may well be proud of this native artist son, who has created such notable productions for one of her public buildings. When placed in their final position in the rotunda of the State College library, they will constitute one of the major reasons why North Carolinians and others should visit Raleigh and Raleigh's leading educational institutions."[8]

When the paintings were actually hung in the college library later that month, it became clear at once that Wells's opinion was not universally shared. L. Polk Denmark, the secretary of the alumni association at the college, immediately wrote a letter to a local newspaper in which he laid down the gauntlet, asserting that such "fantastic specimens of modernistic art" should not be placed in a "beautiful building of classic design." Privately, Denmark maintained that the artist McLean must have the mentality of a two- or three-year-old child. The student newspaper took up the case, editorializing against the paintings with assertions that they "would be more appropriate in the basement of some of our buildings" and that this opinion was shared by "several prominent townspeople who have studied art." A student reporter also interviewed several professors concerning the paintings, reporting comments which ranged from highly favorable statements to remarks that they were totally unsuitable. Architect Jehu D. Paulson thought they were not appropriate in color, scale, or spirit; A. F. Greaves-Walker, a ceramic engineer, found them "so crude they scream at one"; and Alvin M. Fountain from English found his sense of fitness offended because the "gaudy colors" were "unforgivable." By contrast, botanist D. B. Anderson, a member of the library committee whose interest might therefore be said to be vested, considered them a "most desirable addition," which the college was fortunate to have, because they were "well-designed," "vigorous in style," and avoided "both extreme modern and well-worn classical treatment." Mathematician J. D. Clark backed away from the issue, saying that "laymen like me shouldn't judge": evaluation should be left to artists.[9]

The debate became a public matter as well as a campus one. In addition to the letter from the alumni secretary, a newspaper printed one of the student editorials and a letter from Isabelle Bowen Henderson, another artist. Henderson found the paintings to be "agreeably individual in color and style" and "excellent and interesting murals." In summarizing events of the year 1934–35, the historian of North Carolina State College reported merely that "the mural paintings for the library by James A. McLean suddenly made the College art conscious." But in 1939 the guide to North Carolina produced by the Federal Writers' Project listed the library as a point of interest, noting that "modern murals adorn the rotunda."[10]

In the face of the criticism that boiled up, the college library committee wobbled, claiming that the paintings were hung only temporarily and that the permanent disposition would be decided by the general faculty. The latter group responded by pushing the matter again into the hands of the committee, and for several years the question was tossed back and forth. Finally, on 12 September 1941, the library committee presented to the faculty the following resolution: "Recommended that the murals which have been hanging in the rotunda of the library be removed to a more suitable location." A substitute motion was immediately offered, which provided that a group of three outside experts be commissioned to decide if the paintings were "worthy to hang in their present location." After heated debate in which Wells argued for it vehemently, this substitute was defeated by a vote of ninety-three to fifty-six. The original resolution was then adopted. The paintings were removed from the library on 22 September and stored pending their hanging "in a building of modernistic design," probably the next agriculture building for which money had already been appropriated. In commenting on the situation, a local newspaper columnist noted acidly: "A handful of deeply-concerned 'art critics' on the rolling campus of State College apparently have their esthetic noses turned wistfully toward the Art of the middle Eighteenth Century"; he concluded his sarcastic commentary with a plea for contemporary art and living artists.[11]

Plans for the "modernistic" agriculture structure were disrupted by World War II. The murals were stored in the basements of several college buildings and eventually disappeared in a manner still unexplained. About twenty-five years later, McLean's daughter discovered one of the four panels, that devoted to engineering, in a storeroom of the Raleigh Little Theater, where it was used to cover a piece of machinery; the artist then took it into his own possession. Finally, alerted by a person interested in little-known aspects of the history of North Carolina State University, newspapers in Raleigh and Fayetteville in 1982 published stories about the lost murals. In the face of such publicity, officials at the university decided that the time had come for the situation to be rectified. The remaining panel, restored by the artist, was hung in a prominent position in the lobby of the student center on the campus, and McLean was given a ceremonial dinner and considerable attention. By this time, Wells had died so he was never able to savor the eventual triumph of his view of the works. The incident restored McLean to local notice, and in 1985 the Raleigh Little Theater honored him with an exhibition of his paintings.[12]

Wells's wife, Edna, also found Raleigh an interesting place to live, one in which she could find expression for her abilities and interests. At Kansas State Agricultural College she had prepared to be a schoolteacher of science, and on arriving in North Carolina she joined the faculty of the Raleigh High School, where she pioneered in the teaching of biology. She taught for seventeen years and was eventually head of the science department at Broughton High School. She was a serious professional: she soon joined the North Carolina Academy of Science and was an active member of that organization thereafter, contributing significantly to its committee on high school science. In 1920 it was she who discovered the unusually long stem of a princess-tree which Wells thought remarkable enough to describe in his publication "A Phenomenal Shoot." Her professional participation in a cooperative arrangement with Teachers College of Columbia University led directly to Wells's involvement with the New College Community and indirectly, therefore, to his interest in and research on the mountain grassy balds. In the 1930s she undertook graduate study at the University of North Carolina, receiving the M.S. degree in 1936. She then entered upon a program of study that would have led to a Ph.D. degree in botany had not personal tragedy intervened.[13]

Wells and his wife were an uncommon pair in the Raleigh of their day, for there and then it was unusual for a married couple to pursue two separate careers. But they possessed the requisite qualities of heart and mind, and their marriage of more than twenty years was a successful one. Edna's abilities were such that she aspired to the same professional stature Wells himself achieved; this ambition may have introduced into the marriage a mildly disturbing note, for Wells did not enjoy close competition. He was devoted to Edna, however, and the fact that the two shared similar views on important issues gave their relationship stability and satisfaction.[14]

In his relationship with his wife, Wells put into practice views on marriage and morals which he held in principle. He believed that women should find equal opportunity with men in the world of work and rise or not solely on their merits. At the same time, his view of men and women in the home and family was traditional: a male should be the head of his household and hold the right of decision in family affairs; he could, of course, as Wells often did, defer to his wife, but that was a matter of choice. His views on marriage were liberal; once around 1930 he suggested publicly that college students might in many cases be better students if they married, their parents continuing to support or assist them. This opinion brought a storm of criticism down upon him. Even many students at his college were upset at such a radical idea; this was in an age when an instructor at the college could be and was dismissed because he and his wife divorced. Yet Wells openly subscribed to the view that when a marriage was genuinely not working, divorce was an acceptable solution, provided there were no children. In the matter of children, Wells's views were vehement. Children were to be loved, and if they could not be, then they should not be had; consequently, he publicly advocated birth control long before it was acceptable to do so. He believed that under no circumstances should parents, especially fathers, fail in even the slightest degree to meet their responsibilities toward their children.[15]

Wells's fondness for children was immense. It was part of his more general fascination with people in general and his ability to interact as naturally with strangers as with friends. He was never condescending in manner to anyone and therefore communicated easily with young people. Throughout his life he delighted in instructing them in physical skills such as ice skating, or spellbinding them by magically changing pebbles into candy, or awing them by riding a bicycle backward. He loved to associate with them, and they could sense that his feeling was genuine.[16]

Wells's love of young people caused him to behave like a surrogate father or older brother in several cases. In the year 1925–26, for example, two younger botanists on his staff, A. C. Martin and D. B. Anderson, roomed together in his home, the latter testifying later that this was in many ways a familylike arrangement. Later L. A. Whitford and Archie McFarland Woodside, the first two undergraduate biology majors at North Carolina State College, also lived in the Wells home. The help and encouragement which Wells tendered to young students of botany has already been alluded to in the cases of Martin, Whitford, G. F. Papenfuss, R. K. Godfrey, and S. G. Boyce. None of these men might have been motivated to achieve what they did professionally without Wells's stimulation and aid over and above the usual duties of a professor, and all of them have expressed gratitude for his efforts. Edna Wells shared her husband's interest in encouraging young persons of ability. At least two students whom she encountered as a teacher were not only encouraged by Wells to rise to their potentialities but also inspired by him to pursue higher educations. One of these, the unrelated namesake Warner L. Wells, became a prominent professor of surgery in the medical school of the University of North Carolina. The other, Frank Harris Johnson, became a professionally well-known professor of biology at Princeton University. Looking back on his life after his retirement many years later, Johnson cited Wells's attention and friendship as so influential in his development that he was "personally indebted to Bert for more than will ever be known."[17]

Wells's concern for people caused him to be in the forefront of liberal thought on many issues, although his public liberalism was mixed with a personal conservatism in a most unusual way. Politically he was thoroughly democratic by inclination and Democratic by party. With the expansive optimism characteristic of his status and era, he believed in educational opportunity for everyone and that the principal ingredient of success was hard work aimed at realizing native abilities. His early experience with extreme poverty and the greed of landlords in the slums of Toledo made him distrustful of man's quest for money and love of power so he was afraid of commercial monopoly and governmental restriction of liberty. For example, although he used little alcohol himself, he was strongly opposed to prohibition, viewing it as a repressive measure. He was in 1933 one of the founding members of the North Carolina State College chapter of the American Association of University Professors, an organization considered by many at that time to be very liberal, even radical. In the matter of race he held views that did not harmonize with those current in the South of his day. He had attended school with black children in Ohio, and he believed it shameful that children of different races were educationally segregated because he saw that this practice led to schools of low quality. He was also strongly opposed to insults such as the seating arrangements in buses, theaters, and other public places. Privately he attempted to treat everyone as a person, without bias or condescension. For a time he helped a poorly educated black janitor at the college, who was also a clergyman, with the writing and preparation of sermons; he did this as a matter of assisting his fellow human, despite his own views of the organized church.[18]

In the above and other respects the mature personality which Wells displayed during his Raleigh years contained an unusual mixture of attributes. He felt strongly and expressed his feelings emphatically, sometimes without the need for language. He was usually perceptive of and sensitive to the needs of others and yet often overreacted to their words and actions. He possessed dignity without sham, maintaining a distance that was nevertheless open and natural. He was quick to react and slow to change. He could be physically gentle and vigorous at the same time. He could dogmatize and in perfect sincerity take offense at the dogmatism of others. He was a leader who enjoyed directing others and yet could inhabit alone deep areas within himself. He was outwardly expressive and inwardly needful of interaction with the thoughts and ideas of others. His feeling for others was genuine; it was not conditional upon receipt of anything in return. D. B. Anderson, his associate of many years, described him as "a remarkable man, completely unselfish and the most generous human being I have ever known."[19]

Probably as an accompaniment to his sensitivity, Wells found it difficult to accept criticism, even when it was impersonal and dispassionate. He took any negative comment on his scientific work as a personal attack. For example, when H. J. Oosting and others moved to test his conclusions on the salt-spray effect, he was incensed, believing that they were motivated by a desire to discredit him. Like many creative scientists, he was both competitive and territorially possessive; this characteristic made it hard for him to endure honest difference of opinion when it concerned his personal research. When Murray F. Buell, one of his faculty and also a capable ecologist, came to hold views different from his own concerning processes occurring in the Carolina Bays, he could accept the situation, but relations were strained. So when Buell received an offer of a better position at another institution, Wells made no effort to retain him even though Buell did not wish to leave. In science, especially, Wells was at heart a lone worker and could function as part of a research group only if he was the unquestioned leader.[20]

It took all of Wells's strength of personality to deal with a private tragedy that struck in 1938. His wife, Edna, seemed to be at the peak of her life, successful in her work and eagerly pursuing her graduate study. Suddenly she was stricken with cancer; the disease developed quickly, and she died on 6 February 1938. In accordance with both her views and Wells's, a short memorial ceremony was conducted in the company of a few friends; no funeral service of the customary sort was conducted, and no clergyman was present. Instead, Z. P. Metcalf led the proceedings, which included a reading of William Cullen Bryant's poem "Thanatopsis." This procedure caused much local murmuring, as did the cremation of her remains, which was at the time rather new to Raleigh. But she was mourned in the city. The Broughton High School flew its flag at half-staff and held a school-wide assembly in respect to her. Student William Craven and other members of the high school science club went before the mayor and commissioners of Raleigh and induced them to set aside in her memory the Edna Metz Wells Park. This was a small wooded tract near both the Wells home and the school; she had often used it as a source of specimens and for field trips with her classes. Little or nothing was done to the site, however, until many years later some of those same former students pressured the authorities to put it in order. They did, and the park was formally dedicated on 27 May 1978 as part of the fortieth reunion of the high school class of 1938. Although he was then very advanced in years, both Wells and his second wife, Maude, attended the ceremony of dedication.[21]

Wells was a man for whom the married state was the natural one, and he was never a person to withdraw from life, so he married a second time. The woman he chose, Maude Rhodes Barnes, was a perfect mate for him. She was born 11 March 1906 in Clinton, North Carolina, the only daughter in a family of three children. Her parents were Albert Sidney Barnes, a clergyman, who in 1914 became superintendent of the Methodist orphanage in Raleigh, and Daisy Speight Barnes, an artist. After graduating in 1925 from Peace College, a two-year institution in Raleigh, she taught for five years in various elementary schools. The School of Education at North Carolina State College, created in 1927, was open to women, and she enrolled there in 1930. She received the degree of B.S. in high school teaching in 1932 and continued on for one year of graduate study in educational psychology. She then pursued a career in social work. Her first position in this field involved simultaneous service to both Wake County and the city of Raleigh. Her duties were mostly those of a probation officer and welfare worker for white females, but in addition she acted for the police department as a social worker for female prisoners. Later she was able to divest herself of such police activities and served full-time as a probation officer for the juvenile court.[22]

Barnes first encountered Wells when he lectured to a botany class in which she was a student, but she came to know him personally only later as a result of friendship with his wife, Edna. After Edna's death, Wells remained in occasional touch with Barnes; eventually he began to invite her for social engagements, and later it became clear that he was interested in serious involvement. These overtures she routinely declined. Eventually, however, in 1940, she admitted that he was the man for her. Subsequent events proved both of them right.[23]

On 25 February 1941 Wells and Barnes were formally married at the home of her parents before a small group of friends in a ceremony conducted by her father. Because the college was in session, the couple planned only a brief honeymoon stay of a few days at the coast of North Carolina. During the following summer, they made an extended automobile excursion through the Southwest and Mexico; while in Mexico City they stayed in the United States embassy, where Josephus Daniels, a relative of Barnes's, was ambassador. With marriage, Barnes left her professional employment to devote her time primarily to homemaking. During the labor shortage of World War II, however, she returned to welfare work for a year. After the war, she served for two years as executive secretary of the State Legislative Council, an organization that represented a number of women's groups; in this capacity she played an important part in convincing the legislature to establish the first state training school for black girls. Subsequently she worked for a time as a program analyst for the North Carolina Department of Public Welfare and also in the library of North Carolina State College. But her primary aim was to support and contribute to Wells's life, and this she did well. Wells in turn was devoted to Barnes. For example, when he concluded in his ecological study of the Holly Shelter Bay that a freshwater lake had at one time occupied the area, he named that feature "Lake Maude." The marriage was a happy one for both partners; it flourished for almost thirty-eight years, until Wells's death. His later serenity and probably his longevity may be attributed in large measure to the depth of the mutual relationship he and Barnes enjoyed.[24]

Wells's many eventful years in Raleigh were largely happy ones, except for the final few. During the period from 1949 to 1954, when he had ceased to be head of his department, he remained active as a teacher and researcher. But during this time, perhaps as a result of the aging process which affects everyone, certain elements of his personality hardened and some emotional tendencies he had always possessed became magnified. He remained his old many-faceted self with a quick mind, a vigorous body, and an admirable set of personal values; but his short temper contracted even more, his bent toward the formation of snap judgments grew, his tolerance of well-intentioned criticism lessened, and he became less able to admit that he might be wrong in matters about which he felt strongly. The razor-edged objectivity he had wielded so skillfully in the past dulled as a result. His acceptance of the meteorite hypothesis of the origin of the Carolina Bays required him to ignore the main body of geological evidence. Even if the idea should someday turn out to be correct, Wells in his time was acting dogmatically. He had always been inclined to be swept up in his hypotheses, and as he aged those hypotheses more quickly made the transition to fact as far as he was concerned. The catastrophic meteorite hypothesis appealed to his inner proclivities, and now he could become emotional in discussing it. He even altered his view of his own field observations. When he found pollen of aquatic plants in the bottom peat of some of the bays, he wrote in 1949 with the caution of the careful scientist that "while this evidence is definitely against the solution theory, it cannot be interpreted as direct positive evidence for any other theory." By 1953, however, he was citing the same observations as "positive evidence for the sudden or catastrophic formation of the depressions." And much later, in 1971, he maintained in a letter to another scientist that "the proof of the meteorite origin of these Carolina Bays lies in the fact that in the basal peat the pollen is that of aquatic plants. . . . Only the meteorite theory can support this fact." He thus disregarded the considerable body of scientific evidence which did not harmonize with his chosen explanation. Such a change in interpretation might be taken as simply the result of increased conviction, but his diminished objectivity was manifested in other cases as well. After he retired, for example, he embraced the idea that the offshore islands (outer banks) of North Carolina are the result of sand spits built up by wind and wave action. This notion was not original with him; it was an old concept that never gained acceptance. Wells thought it self-evidently true, and when he discussed it in a seminar for botanists, he presented a completely one-sided account that omitted mention of opposing evidence and opinion. This performance saddened listeners who knew of his lifetime of outstanding accomplishments.[25]

After Wells relinquished its headship, his academic department fell upon troubled times. New faces and new forces were at work in his old world. It would have been difficult for a lesser man to absorb such changes quickly and refrain from overinvolvement in them. For Wells, after thirty years of exercising strong, stimulating, and unquestioned leadership of a vigorous personal sort, it was impossible. When two of the new young botanists became dissatisfied with the administrative practices of his successors and actively fomented discord, Wells unfortunately allowed himself to be drawn into the strife on their side. The result was division, a souring of personal relationships, and unhappy days. It became a time of uneasy acceptance and wary toleration of an unpleasant situation for all concerned. In any organization a considerable readjustment usually follows the passing of such a strong and dynamic leader as Wells. In this case it was not finally completed until all three dissenters were gone, Wells by his complete retirement in 1954.[26]

Scrutinizing the development of weakness in a vigorous and outstanding man is painful. But objectivity requires it: Wells at the close of his career was no more immune than anyone to human frailty. Fortunately, in their fullness and serenity his final years were a more fitting conclusion to an eventful and productive life.

CHAPTER 14

Retirement Years, 1954–1978

I remember the day we came out here, so long ago. . . . We were just carried away with it. We climbed all over those rocks down there. So we would come back often and bring a picnic. . . . Well, we wanted a little place, just a weekend place, so we kept looking. One day we were going to Durham to get a . . . washing machine. He got a hundred dollars out of the bank and gave it to me for it. On the way over to Durham I said, "Let's get a newspaper and see if there's anything over here for sale." . . . It was lunch time, so we went there first. . . . She brought our lunch back, and . . . a Durham paper. I opened it up, and [there] was a place for sale . . . on the Neuse River. The minute I read it I knew exactly what and where it was. . . . I handed it to Bert, and he read it, and he turned absolutely white. He knew immediately, too. . . . We got up and left that food. . . . We walked in there and Bert said, "Maude, give me that hundred dollars back," and plunked it down.

—Maude Barnes Wells, 1979

WELLS FORMALLY retired from his professional position at North Carolina State College in 1954. But years before this he felt a need for a home that was closer to nature than his longtime residence at 1605 Park Drive in Raleigh. Shortly after his wife, Edna, died in 1938, he purchased a cottage at Carolina Beach, North Carolina; he used this house not only for vacations but also as a base for summer research and numerous field trips. Some years later he sold this cottage and bought the Stuart House, a beautiful old building on the bank of the Cape Fear River at Southport, North Carolina. This large house was a historic one, having been operated as a well-known inn for nearly one hundred years, from 1834 to 1929. He used this house in the same way he had employed the beach cottage until it was destroyed by a hurricane in 1954. By that time, however, he had acquired another place which wholly determined the character of his final years.[1]

Wells and his wife, Maude, had made a habit in the 1940s of taking Sunday excursions through the countryside around Raleigh, looking for possible remnants of the original natural vegetation of the area. On one such occasion they investigated a site which Wells had chosen from a map as a likely spot. About twenty miles from Raleigh and near the town of Wake Forest, the 144-acre tract was a run-down old farm; it included considerable woodland and an extremely scenic section known as Turkey Neck, where the Neuse River reversed itself twice in a double bend. Here the stream flowed past granitic outcrops and bluffs, topped at the end of Bent Road by the fifty-foot-tall overhanging feature called Ziegle's Rock. Wells and Maude fell in love with this place, returning to it many times, often with picnic lunches, down to the rocks, those beautiful rocks. They eventually met the elderly owner and toured the property, which included a primitive old house and dilapidated farm buildings. They had been thinking of getting a little weekend place, and Wells often spoke of asking the owner of the farm to sell him an acre or two. But he never did.[2]

Then on a day in December 1950, one that they later called "the most fantastic day of our whole lives," Wells and Maude made a fateful trip to the Sears, Roebuck store in Durham to purchase a new washing machine. Opposite the store they stopped for lunch. Maude felt easy, for the $100 Wells had extracted from the bank was safe in her purse, ready to fulfill its purpose. Although the plain lunch before them invited, she paused to open the newspaper the waitress had brought and casually scan the real estate section. Suddenly excitement surged through her. It had to be. Although no specific location was given, she knew at once what site the advertisement described. There, defined in a box on the crowded page, was their beloved spot on the rocks beside the Neuse: 144 acres, equal distance to Durham and Raleigh, in loop of river, so-and-so many buildings, electricity, such-and-such timber and land, ideal for a gun club. They had never counted the buildings or known of the electricity, but it was their place. So. Without a word she handed the paper to Wells, who read it and blanched. He knew at once, too. After a silent minute, he strode to the telephone. Learning that the farm had not yet been sold, they left their lunch uneaten. That day, with typical forcefulness, Wells bought the place, using the washing machine money as a binding deposit. "It was," Maude said later, "the most wonderful thing that ever happened to us." Wells previously had thought of retiring to the coast when the time finally came, but now he determined that it would be to this old farm, which embraced such natural beauty in its quiet isolation.[3]

The site was beautiful, but making the farm a comfortable retirement home would require effort daunting to most persons. Living conditions would be primitive at first. The old house with its low-ceilinged rooms was in poor repair. There was electricity, but heating required burning wood. Water had to be dipped from a well, baths and showers taken from a bucket. A mile separated the mailbox from the front door. But Wells knew that when he retired he would be forced to leave the life of science largely behind: he would never be able to keep up with its increasing complexity and specialization, and the process of aging would inevitably reduce his faculties. At the same time, he could never desist from active life until physical infirmity forced him to; both he and Maude were what she called "busy people." The farm provided what was to Wells a perfect solution. Work would make it a home to enjoy, and the more work he could do the more enjoyment there would be.[4]

Some of his friends thought he was mistaken, to say the least, to retire to such a place; few could see its possibilities. But Wells did. He essentially took on a new career, a private one fully supported by

In his studio at Rockcliff Farm, B. W. Wells, after his retirement in 1954, with one of his paintings of Ziegle's Rock and the Neuse River. Courtesy of Maude Barnes Wells.

Maude, in which through years of hard work he made that old farm match the natural beauty of its site and fit his retirement life perfectly. He repaired and remodeled the house, installed a pump and water line, constructed a spring house and pump house, restored two old erosion-control terraces, cleared land for scenic views, inventoried plant species, and laid out ten miles of nature trails and footpaths. And every day he did the chores inherent in an isolated existence. He built from scratch a building to serve him as a combination office, workshop, and art studio, complete with skylight. He found on the land a tremendous pile of rocks of various sizes and used these to construct two features of which he was ever proud. In his studio he fashioned a large stone fireplace and chimney, using a six-foot half of a tree trunk as a mantel. And outside he built a long, low stone wall to set off a row of cultivated wildflowers. In the end, his efforts made the old farm a place of honest rustic beauty. It became, in his words, "a very, very desirable place to continue to live."[5]

Continue to live he did. As the work involved in developing the farm gradually diminished with accomplishment, another aspect of Wells's retirement career grew to the dimensions of an almost full-time occupation. This was painting. He had long been interested in art and had from time to time taken steps to improve his proficiency. Now in his retirement he blossomed into an active amateur painter. He toiled away almost every day in the studio he had built and took another correspondence course. Working mostly in oil, but also in crayon, watercolor, and acrylic, he created hundreds of pictures. The exact number of his works is unknown, but one partial inventory which he made in 1973 listed 342 items. The subjects on this list fell into the following categories: landscapes (202), portraits (82), views of buildings (48), and still lifes (10); he painted no abstracts. He gave away most of his paintings to friends and acquaintances. In this regard he was a soft touch. As he himself put it in 1971, "Somebody comes in here and says, 'Oh, I like this painting,' and I'll take it off the wall and say, 'Well, you just keep it.' That is a very easy way, you know, to get rid of something."[6]

For Wells, of course, the process of painting was the important thing, not the possession of the finished results, although one should not discount his pleasure at the thought of acquaintances viewing his works frequently in their homes and remembering him through them. He had chosen an activity for his old age well, for, as he put it, "In art one can improve. . . . In art your memory does not fail, you go on improving in art. But in science your memory fails because you never can remember all those scientific terms. This is a very striking difference between art and science." He spoke truly, for by his last years he was able to remember almost no plant names or scientific terms. In his painting he pushed his ability seriously and hard, and he made himself a talented artist. He did not merely repeat techniques he had already largely mastered but continually tried new approaches, for even in old age he was, as he said, "one of these chaps that enjoys trying to do things that are new, somewhat different." He never claimed to be as capable as a professional; in fact, he often disclaimed any such pretension. Still, he was proud of his accomplishment: in one summary of biographical data he listed himself as a "gifted amateur artist."[7]

With all the physical labor and creative activity of farm work and painting, Wells's days were full indeed. His attitude is summed up in some lines from "The Rockcliff Farm—A Letter to Children," a piece he once penned for his nieces:

With science and art and all the farm chores
The days are too short and one only deplores
The need for sleep, for it's lots more fun
To be active than lazy and feel outdone.

Active he certainly was. And through it all, until very near the end, his intellect still hummed as he busied himself in his office on a regular daily basis. He composed articles for popular consumption. He also wrote numerous letters to friends describing his life, letters to scientists about their researches, and letters to newspapers on public issues. Once in response to an article on Cades Cove published by Supreme Court justice William O. Douglas, he sent the justice a discussion of the grassy balds of the mountains, along with several of his publications. Douglas read these with interest and expressed regrets that he had not known about them, for they "would have enriched" his article. He continued to read as much as ever, especially about national and world affairs. Even at the end of his life, when the number had dwindled considerably, he still subscribed to more than twenty-five magazines and the daily *New York Times*. When he was not actively moving or working with his hands, he was usually reading; he himself put it that he "just could not do a plain sit." He was also always eager to discuss all that he read. One interviewer found that at age ninety-three Wells still exhibited an enthusiasm that was infectious; another reported that he was still a teacher, beginning the interview with a brief field trip and lecture.[8]

The farm became, in effect, a privately owned nature preserve. Wells and his wife enjoyed sharing it with others, and many people visited them there, friends and strangers, individuals and groups. His presence made the farm a place to which they were drawn, and his imagination and hard work had made it a place the beauty of which they thoroughly enjoyed. After visiting, many of them later wrote to express their appreciation in glowing terms. One such letter, for example, said, "It is inspiring to be with those interested in conserving the marvelous beauties of nature. If I hadn't had it before, I feel that I would have caught a love of the things growing from your enthusiasm." But no one's love of the farm was greater than his own. For the twenty-four years of his retirement it gave him a pleasure, a peace, a serenity few enjoy. How he felt at the end of a day's work, when he sat outside in the evening with a glass of wine, contemplating meadows and trees and sky, is revealed in his verse-letter for children:

> You come to the end of a dead-end road,
> And, crossing the line at a dip, your load
> Of city worries, cares, and thought
> Drops as you cross into Rockcliff Farm
> Where peace, freedom, and joy were bought
> With the acres.

At such moments he truly felt himself to be what he often claimed: the richest man in the world.[9]

Wells's health remained remarkably good, and he had the fortune to be busy and active almost to the end. As he aged, it was hard for him to think of himself as old. Once, on the occasion of a meeting of the Wild Flower Preservation Society at the farm, he pulled a much younger friend aside and charged him vehemently: "I want you to go along on that hike, and I want you to stay up close to the front! You've got to look out for all those *old* people!" And yet, as he did grow old and death did near, he made his plans and expressed his wishes. Before he and Maude undertook an airplane journey in 1968, he wrote a letter to a friend directing him in the event of a crash "to carry out the cremation procedure for both bodies if they can be recovered. There will be no funeral." At that time he wanted his ashes to be scattered at Ziegle's Rock; always the ecologist, he added in the margin of this letter the following comment: "Note: ashes do not pollute." He made his plans, but he never talked much or worried about dying. He always hoped that he would merely fade out at the end, and he did just that. He experienced some severe digestive trouble and entered a hospital for tests; there he contracted pneumonia, a disease his worn-out body could not withstand. On 29 December 1978, at the age of ninety-four, he died. The next day friends gathered to pay their respects, and the day after that a private memorial ceremony was held in his honor by a few of them. He was gone.[10]

Wells was a man deserving of honors. He was primarily a plant ecologist; yet he did not receive, and still has not been accorded, the national recognition he deserved for his contributions to that science. Some possible reasons for this state of affairs have already been suggested. In 1979, however, two brief obituary notices in scientific publications called attention to his life, and in 1986 the botanical community was provided with a detailed study of his ecological work. But more locally, in the state of North Carolina and at North Carolina State College, his contributions were widely recognized, and during his active career he received several public acknowledgments of them. In 1932 he was given an honorary life membership in the Garden Club of North Carolina. Probably the first such award bestowed by that organization, this honor was in recognition of his creation of *The Natural Gardens of North Carolina*. On his retirement in 1954, the dean of the School of Agriculture at North Carolina State College commended him on his long service to that institution. Wells's record of performance, the dean wrote, was that "of a productive scientist and an enthusiastic person" and "an important part of the history of this institution." In 1955 he was named "Tar Heel of the week" by a Raleigh newspaper and recognized with an account of his life and works. Congratulating him on this latter notice, a longtime friend wrote, "Of course you have been a Tar Heel for many years; you have added much to the state and brought out many ideas and expert knowledge to make it what it has become."[11]

North Carolina State College also honored him in retirement. In February 1963 Wells learned that he would be awarded the honorary degree of Doctor of Science, which was conferred on 1 June. The citation that accompanied this honor mentioned the breadth of his interests, the warmth of his personality, and his long service to the college in teaching, research, and administration. But it particularly emphasized his dedication to and defense of the ideals of science and a university: "Because you realized that scholarship cannot remain cloistered when its essential interests are threatened, you led the fight to prohibit political interference in the curriculum and thereby helped to establish an essential ingredient of academic freedom—the right and the duty of the scholar to lead the community to a reliance on information painstakingly acquired and honestly presented." In receiving this honor at that particular time, Wells was in a sense a member of the last class to be awarded degrees from North Carolina State College, for the name of the institution was changed as of the following July. That name had originally been adopted in 1917, and Wells's long association with the school closely matched the period during which it was known by the name most familiar to him.[12]

Wells's old Department of Botany did not forget him either. Even though all the faculty members with whom he had worked had departed or retired, it honored him in March 1974 on the occasion of his ninetieth birthday. This recognition took the form of a week-long public exhibition of a large number of his paintings in the student center on the university campus; the celebration was also marked by a reception attended by several hundred of his friends and admirers. Later, in October 1977, the department staged a special commemorative dinner to honor together five living men who had served as its head. As the doyen of the honorees, Wells was the center of attention at this event.[13]

After his death, honors to Wells multiplied. Although these acts of recognition were performed too late for him to derive personal pleasure from them, they were nevertheless richly deserved. On 18 May 1979 the board of trustees of North Carolina State University adopted a resolution honoring his memory. In this measure the board noted his long service to the university, his research on insect galls and ecology, and his stand before the legislature on evolution; it therefore unanimously acknowledged his "decades of service . . . to the University, as a teacher and scientist and as one devoted to freedom of inquiry and enlightenment." On 28 September 1979 the North Carolina Wildlife Federation presented him posthumously with its Governor's Award in Conservation Education. The organization asserted that "probably no one has done more to increase our appreciation of the beauties of nature and to foster conservation of plant life in the Southeast" and cited him as "one of the most beloved and influential educators in North Carolina." On 9 March 1981 one of the large halls in a new biological sciences building at North Carolina State University was dedicated as the Bertram Whittier Wells Auditorium; this memorial action was appropriate for one who had regarded himself as primarily a teacher. And in 1985 a new kind of ornamental creeping blueberry was developed by plant scientists at North Carolina State University; this cultivated plant was named 'Wells Delight' in his honor. Nor was his art neglected. Two of his paintings are now held in the permanent collection of North Carolina State University, and in 1989 a tribute to his impact on conservation and nature in North Carolina reproduced two of his landscapes in full color.[14]

Wells's beloved retirement home, Rockcliff Farm, may serve as a continuing monument to his memory and spirit. In 1965, long before he died, Congress approved plans by the United States Army Corps of Engineers for a dam on the Neuse River; the resulting impoundment would inundate much of Wells's property, although leaving the homesite intact. He therefore sold the tract but retained the right of lifetime residence on it. As construction of the dam approached, various persons and groups pressed for preservation of the farm site as a natural area devoted to the causes of nature study and conservation which were so dear to Wells's heart. These interested parties included the Wake and New Hope Audubon chapters, the North Carolina Botanical Garden, and the Conservation Council of North Carolina. The dam was completed and its reservoir filled in 1983. To the Division of Parks and Recreation of the state of North Carolina the Corps of Engineers entrusted the management of the remaining tract as the B. W. Wells Interpretive Center. Maude left the farm in 1981. Soon after Wells's death some members of the Raleigh–Wake County Audubon Society had begun working voluntarily on the farm to keep it in good condition. Now they and other friends and admirers joined in forming the B. W. Wells Association. Formally incorporated in 1982, this organization was founded to assist the state in developing and maintaining the Wells farm as an area for the study of living organisms in their natural environments. The association and its approximately one hundred members have continued to press for the realization of this goal through publicity, the political process, and individual exertion. Pending completion of development of the site, a small cadre of its members have been especially devoted in contributing their time and hard physical labor. As the president of the association put it in 1986, "We're a corps of about a dozen people. We paint old buildings. We rake leaves. We dig and weed flower beds. We're holding the fort until something happens." Their efforts have borne real fruit, for they have restored and developed trails, landscaped the farmhouse area, removed a decrepit barn, established new plantings of native wildflowers, and carried out maintenance work on the grounds and remaining buildings. Both the Wells Association and his old farm site are honors and memorials befitting the spirit expounded for so many years by Bertram Whittier Wells.[15]

EPILOGUE

So who was Bertram Whittier Wells?

He was a remarkable man. He was an outstanding scientist, a creative pioneer ecologist. He was a forceful champion of nature, an apostle of natural beauty for all. He was a gifted teacher and mentor, a preceptor of the way of intellect. He was a fierce fighter for academic excellence, a stout defender of truth, and a resolute foe of error and deception. He was an able artist. He was a gentle lover of people and a steadfast friend. He was a second father to children not his own and an advocate of the right of all children to learn everything. He was a clear example of life with vigor and death with dignity. He was the best Bertram Whittier Wells that he could be.

His life carries a message. For Wells was an effective human being, who, despite his limitations and imperfections, used his particular abilities to the utmost. His life shows how one individual can contribute positively to the world by the full exercise of personal talents, whatever those talents may be and whatever the dimensions in which they operate. It shows the personal and public value of respecting emotion but using intellect, of following the reasoning mind wherever it leads. It demonstrates that to the extent of their abilities all who truly wish it can know books and thoughts, use eyes and brains to examine the world firsthand, and commit creative acts that truly make a difference.

Most of all, his life exhorts us to be careful. When he was born, Chester Arthur occupied the White House and the federal Union contained thirty-eight states; Queen Victoria reigned over a still-growing empire and Benjamin Disraeli was her prime minister; the internal combustion engine was a novelty; and radio, television, airplanes, and space exploration were unknown. By the time he died, the world had experienced two tremendous wars, total political restructuring, and a technological explosion producing rapid global transport, instant electronic communication, and the possibility of nuclear holocaust. Science and technology had given unprecedented human control to the realms in which they rule. But science had also brought the realization that our earth is not infinite, our resources not unbounded, our understanding of complex interaction not yet complete. Wells adapted to the stupendous changes that occurred during his lifetime as well as anyone, but he also absorbed better than most the lesson of limitation and interrelation.

So his championing of nature directs us. He spoke mostly in North Carolina, but his message still sounds for the whole world. Make progress, it says, for the betterment of humankind, but respect the meshwork of our earth. Stride purposefully ahead, but stride softly: hear the trees, the grass, the flowers; read the land, the air, the waters; honor all nature, of which we are but part.

NOTES

Chapter 1

1. Iden, 1926. O'Keef, 1955. Wells, F1967.
2. Wells and Shunk, A1928b.
3. McIntosh, 1976; 1985. Shimwell, 1971. Tobey, 1981.
4. McIntosh, 1976; 1985. Shimwell, 1971. Tobey, 1981.
5. Egler, 1951. McIntosh, 1976. Troyer, 1986.
6. Cooper, 1979a; 1979b. Egler, 1977, p. 15.
7. Crafton and Wells, A1934. Wells and Shunk, A1928b; A1931; A1938c. Wells, A1946a.
8. Wells and Shunk, A1928b; A1931; A1938c. Wells and Boyce, A1953c.
9. Egler, 1977, p. 14. Tobey, 1981. Whittaker, 1953; 1962. Shimwell, 1971. McIntosh, 1976; 1985. Clements, 1916; 1936.
10. *North Carolina Agriculture and Industry*, 1925, 15 January; 1926, 27 May. Wells, A1928a; A1939; A1942; A1946a. Wells and Shunk, A1928b; A1931. Crafton and Wells, A1934.
11. McIntosh, 1976; 1985.
12. Wells, A1942; A1946a. Wells and Shunk, A1931. Troyer, 1986.

Chapter 2

1. Wells, A1924b. Harshberger, 1911. Cooper, 1979a; 1979b.
2. *North Carolina Agriculture and Industry*, 1925, 15 January; 1926, 27 May. Wells and Shunk, A1928b.
3. Personal communication, D. B. Anderson, 11 February 1978. Wells, A1928a; D1952. Wells and Shunk, A1928b.
4. Wells and Shunk, A1931; A1937b; A1938b; A1938c. Wells, A1929a; A1936a; A1936b; A1936c; A1937a; A1938a; A1938d; A1939; A1942; A1946a; A1949. Dachnowski-Stokes and Wells, A1929b. Crafton and Wells, A1934. Wells and Boyce, A1953a; A1953b; A1953c; A1954.
5. Troyer, 1986.
6. Cowdrey, 1983. Troyer, 1986.
7. Wells, A1924a; A1924b; A1928a; F1932/1967. Wells and Shunk, A1928b; A1931.
8. Wells, A1924a; A1924b; A1942. Troyer, 1986.
9. Crow, 1973. Troyer, 1986.
10. Troyer, 1986.
11. Wells, A1924b; A1928a. Wells and Shunk, A1931.
12. Wells and Shunk, A1931. Wells, A1942; A1946a. Wells and Whitford, A1976[1977].
13. Oosting, 1948, p. 255. Gleason and Cronquist, 1964, p. 321. Troyer, 1986.
14. Troyer, 1986.

15. Wells and Shunk, A1928b; A1931. Wells, A1928a; F1932/1967. Crafton, 1932. Crafton and Wells, A1934.

16. Johnson, 1969. Billings, 1970. Troyer, 1986. Golley, 1977.

17. Troyer, 1986.

18. Wells, A1936a; A1936b; A1936c; A1937a; A1938a.

19. Wells, A1936b; A1936c.

20. Wells, A1937a. Troyer, 1986.

21. Wells, A1937a; A1946b. Mark, 1958.

22. Wells, A1936a.

23. Troyer, 1986. Brown, 1941. Wells, A1956; A1961.

24. Troyer, 1986. Gersmehl, 1970. Smathers, 1980.

25. Troyer, 1986.

26. Wells and Shunk, A1931; A1938b; A1938c.

27. Wells and Shunk, A1937b; A1938b; A1938c.

28. Wells, A1938d; A1939; A1942.

29. Wells and Shunk, A1938c. Personal communications, D. B. Anderson, 11 February 1978; L. A. Whitford, 14 July 1979. Oosting and Billings, 1942. Oosting, 1954.

30. Oosting and Billings, 1941; 1942. Personal communications, D. B. Anderson, 11 February 1978; H. T. Scofield, 22 January 1978; L. A. Whitford, 16 February 1978.

31. Bayer, 1938. Oosting and Billings, 1942. Oosting, 1945; 1948; 1954; 1956. Boyce, 1954. Troyer, 1986.

32. Wells, D1938a, p. 21; A1942, p. 558; A1949. Wells and Boyce, A1953c. Troyer, 1986.

33. Troyer, 1986.

34. Wells and Boyce, A1953a; A1953b; A1954.

35. Wells and Boyce, A1953c.

36. Troyer, 1986. Whitehead, 1972.

37. Whitford Papers, Letters, 1962, H. Godwin to H. C. Brown, 9 November; A. W. Bayer to H. C. Brown, 22 November.

38. Personal communication, A. W. Cooper, 27 June 1984. Lawrence, 1979.

39. Personal communication, A. W. Cooper, 27 June 1984.

40. Egler, 1977, p. 3. Snyder, 1980. Lieth, Landolt, and Peet, 1979/1980/1981.

Chapter 3

1. Sears, 1914. Wells, B1914. Cook, 1923.

2. Wells, B1915; B1921a; B1921b. Wells and Metcalf, B1921d. Trotter, 1915. Gagné, 1984.

3. Personal communications, M. B. Wells, 22 September 1979; L. A. Whitford, 21 August 1979. Wells, B1918. National Union Catalog, 1979. Felt, 1918. Wells Papers, J. F. Gates Clarke to B. W. Wells, 7 January 1963.

4. Wells, B1916. Trotter, 1916. Cook, 1917a; 1917b.

5. Wells, B1920. Trotter, 1920. Zweigelt, 1931. Liser y Trelles and Molle, 1945. Smith and Taylor, 1953. Walton, 1960. Dundon, 1962. Lewis and Walton,

1964. Elias, 1970. Pollard, 1973. Kandasamy, 1982. Gagné, 1983. Hodkinson, 1984.

6. Wells, B1921c; B1921e.

7. Trotter, 1922/1923. Kinsey, 1920. di Stefano, 1967; 1968.

8. Cook, 1923. Wells, B1923. Kinsey, 1930. Zweigelt, 1931. Liser y Trelles and Molle, 1945. McCalla, Genthe, and Hovanitz, 1962. Buhr, 1964. Maresquelle and Meyer, 1965.

9. Martin, 1924. Wells, B1934; D1941a. Trotter, 1935. Felt, 1940.

10. *American Men of Science*, 1921, p. 727; 1927, p. 1043; 1933, p. 1187; 1938, p. 1514; 1944, p. 2068; 1949, p. 2662; 1955, p. 1207; 1961, p. 4353. Wells Papers, E. H. Davison to B. W. Wells, 12 July 1965.

11. Wells, C1910. Fritsch, 1945. Papenfuss, 1951. Norton, 1969.

12. Wells, C1919; C1921. Prouty, 1921.

13. Wells and Shunk, A1931. Wells, F1932/1967.

14. Whitford Papers, Notes; Undergraduate Years.

15. Wells, C1928; C1929. Ahles, 1964. Reynolds, 1968. Primack and Wyatt, 1975. Whitford Papers, Notes. Personal communication, S. G. Boyce, 15 September 1981.

16. Wells, A1940; C1943. Wells and Shunk, C1941.

17. Wells, C1944; C1947. Barker and Zullo, 1980. Whitford Papers, Notes. Personal communication, A. W. Cooper, 18 August 1984.

18. Wells and Shunk, A1928b.

19. O'Keef, 1955. Whitford Papers, Notes. Wells, D1927. Wells and Shunk, A1928b; A1931.

20. Wells and Shunk, A1928b. Whitford Papers, Notes. North Carolina Department of Agriculture, 1926–28, p. 47. Wells, A1929a. Dachnowski-Stokes and Wells, A1929b.

21. Darrow, 1960. Galletta, 1975. Wells and Shunk, A1928b.

22. Wells and Shunk, A1931.

23. Coville, 1937. Darrow, 1960. Galletta, 1975. Ballington, 1984. North Carolina Agricultural Experiment Station, 1938. North Carolina Department of Agriculture, 1926–28, p. 56; 1930–32, p. 52; 1932–34, pp. 89–90; 1934–36, p. 81; 1936–38, p. 87; 1938–40, p. 126.

24. Wells and Shunk, A1928b.

Chapter 4

1. Payne, 1977, p. 14. Whitford Papers, Notes.

2. Wells, F1932/1967, pp. 116–17. Wells and Shunk, A1931, p. 487.

3. O'Keef, 1955. Wells and Shunk, A1928b.

4. O'Keef, 1955. Garden Club of North Carolina, 1935, pp. 41–42; 1936, pp. 20, 26–27. Totten, Ballard, and Little, 1959, p. 42. Wells, F1932/1967 (1967 reprint), p. xviii. Baldwin, 1968, pp. 92–93.

5. Garden Club of North Carolina, 1937, pp. 30, 31, 41. Totten, Ballard, and Little, 1959, p. 76. Wells and Shunk, A1928b.

6. Wells, F1939/1955. Federal Writers' Project, 1939, p. 332. Robinson, 1955, p. 347.

7. Wells, D1953. Anonymous, 1951. Wells Papers, 1959, R. H. Pough to Mrs. C. Appleberry, 18 December. White, 1960. Whitford, Notes. Personal communication, D. A. Adams, 29 June 1992.

8. Teale, 1951, pp. 237–38. Wells Papers, 1949, E. W. Teale to R. H. Pough, 5 April; 1959, Pough to Wells, 6 April; Pough to T. W. Morse, 6 April; Pough to L. Ballard, 6 April.

9. Wells Papers, 1957, Wells to Pough, 21 December; 1958, Pough to Wells, 2 January, 12 January. White, 1960.

10. Wells Papers, 1959, Pough to Wells, 6 April. White, 1960.

11. Wells Papers, 1959, Pough to Wells, 6 April, 25 April, 18 December; Pough to L. Ballard, 6 April; Pough to T. W. Morse, 6 April, 25 April; Pough to Mrs. C. Appleberry, 18 December. Personal communication, A. W. Cooper, 29 June 1992. White, 1960. Wells, F1967, p. 14.

12. Wells, F1932/1967, p. 442; F1936/1959. Garden Club of North Carolina, 1933, pp. 24–30.

13. Shelford, 1926, pp. v–vii, 724. Metcalf and Wells, F1926. Ecological Society of America, 1937; 1942, pp. 24, 30.

14. Lawrence, 1966. North Carolina Wild Flower Preservation Society, 1956, 2(4), p. 3; 1957, 3(1), p. 2; 1972, March, p. 7. Wells Papers, 1951, Mrs. Herbert P. Smith to Wells, 24 August; 1956, L. Melvin to B. W. Wells, 24 January; North Carolina Wild Flower Preservation Society letterheads. Wells, C1929.

15. North Carolina Wild Flower Preservation Society, 1957, 3(1), p. 2; 1966, 8(1), pp. 3, 9.

16. Papers of A. W. Cooper, B. W. Wells to Cooper, 19 June 1964, 24 March 1966. Stick, 1985.

17. Wells Papers, 1972, B. W. Wells to the *News and Observer*.

18. North Carolina Nature Conservancy, 1983, No. 23, p. 3; 1985, No. 31, pp. 1, 5.

Chapter 5

1. North Carolina Wild Flower Preservation Society, 1969, 9(2), p. 12.

2. *North Carolina Agriculture and Industry*, 1925, Dr. B. W. Wells to lecture, 22 January. Whitford Papers, Notes. Schneider, 1983.

3. *State College Record*, 1924, Technical and general lectures 1924–25, 23(8), pp. 11–12. *Technician*, 1936, B. W. Wells to speak at Charlotte meeting, 8 May. Whitford Papers, Notes.

4. *Technician*, 1935, Garden enemies listed in speech, 25 October. Garden Club of North Carolina, 1937, pp. 50, 65; 1938, p. 25; 1939, p. 24; 1942, Directory, p. 12. Anonymous, 1951. Totten, Ballard, and Little, 1959, pp. 4, 42.

5. Whitford Papers, Notes. Carpenter, 1978, pp. 40–41. Carpenter and Colvard, 1987, p. 251.

6. North Carolina Wild Flower Preservation Society, 1955, 2(4), pp. 1–2; 1964, 6(4), pp. 2, 6; 1967, 8(2), p. 8.

7. Wells Papers, no date, Record of B. W. Wells. North Carolina Wild Flower Preservation Society, 1964, 6(4), p. 2. Whitford Papers, Notes.

8. Personal communications, C. H. Green, June 1984; J. Downs, 12 July

1984. Overton, 1987. Quillen, 1992. *News and Observer*, 1992, End of a long nature trail, 28 March.

9. Iden, 1912; 1920. Personal communications, P. Talley, 17 June 1986; M. B. Wells, 23 June 1986. Wells, F1932/1967, p. 335.

10. *Technician*, 1935, Garden club invites Wells to lead jaunt, 4 October.

11. Iden, 1926.

12. Wells, D1938a; D1938b.

13. Wells, D1953.

14. Shelley, 1974. Personal communication, R. A. Popham, 16 August 1979. Green, 1974.

Chapter 6

1. Iden, 1926. Wells Papers, no date, Record of B. W. Wells.

2. Robertson, 1924. Gatewood, 1960b, p. 235.

3. Wells, F1923a.

4. Wells, F1923b. Reagan, 1987, pp. 78–79. Carpenter and Colvard, 1987, pp. 211–12.

5. Wells, F1924–25; A1924b.

6. Wells, F1924–25.

7. Wells, F1924–25.

8. *Raleigh Times*, 1924, 22 November. *Technician*, 1924, Favorable comment on articles by State College professor, 28 November.

9. Wells, F[n.d.]. Gatewood, 1960a; 1960b, p. 216.

10. Wells, F1927; F1949; F1950. Metcalf and Wells, F1926.

11. Wells, F1939/1955.

12. Wells, F1944; F1960.

13. Wells, F1961a.

14. Wells, F1956; F1961b; F1967; F1971, p. 11.

15. Personal communications, D. B. Anderson, 18 April 1982; L. A. Whitford, 20 July 1982.

16. Wells, F1932/1967. Small, 1913; 1933. Greene and Blomquist, 1953, p. 137. Justice and Bell, 1968, p. 198. Duncan and Foote, 1975, p. 210. Radford, Ahles, and Bell, 1968.

Chapter 7

1. Wells, F1932/1967.

2. Wells, F1932/1967, pp. 53, 83, 101, 127, 132, 136, 138, 141, 181–82, 202.

3. *Biblical Recorder*, 1932. Daniels, 1932. Anonymous, 1933a. Cromelin, 1933. Fuller, 1933a; 1933b.

4. Federal Writers' Project, 1939, p. 216. McPherson, 1976, pp. 21, 110. Totten, Ballard, and Little, 1959, p. 19. UNC Press Records, 1932, Iden to Couch, 16 September.

5. UNC Press Records, 1932, Iden to Couch, 16 September. Garden Club Archives, 1931, Minutes of the Executive Board, 16 January.

6. UNC Press Records, 1931, Couch to Iden, 16 April, 21 April. Garden

Club Archives, 1931, Minutes of the Executive Board, 22 April. Couch, 1946, p. xiv.

7. Garden Club Archives, 1931, Minutes of the Executive Board, 22 April; Minutes of the annual meeting, 23 April. Totten, Ballard, and Little, 1959, p. 4. UNC Press Records, 1931, Iden to Couch, 29 April; Tomlinson to Couch, 15 May.

8. Garden Club of North Carolina, 1931, p. 18. UNC Press Records, 1931, Iden to Couch, 29 April; Couch to Tomlinson, 5 May; Tomlinson to Couch, 15 May.

9. UNC Press Records, 1931, Tomlinson to Couch, 13 June, 30 July; Couch to Tomlinson, 16 June; 1932, Iden to Couch, 31 March, 16 September; Wells to Couch, 9 June; Wells to W. C. Coker, 30 August.

10. UNC Press Records, 1932, Iden to Couch, 31 March; Tomlinson to Couch, 7 July; Couch to Tomlinson, 9 July.

11. UNC Press Records, 1932, Wells to Couch, 9 June, 14 July; Tomlinson to Couch, 18 July; Iden to Couch, 16 September.

12. UNC Press Records, 1932, Couch to Wells, 9 July, 26 July; Wells to Couch, 14 July, 29 July, 11 August, 3 September; Couch to Tomlinson, 15 July, 19 July; Tomlinson to Couch, 18 July, 22 July.

13. UNC Press Records, 1931, Couch to Iden, 16 April; Iden to Couch, 29 April; Tomlinson to Couch, 15 May, 13 June, 30 July; Couch to Mrs. E. L. Tarkenton, 10 June; 1932, Iden to Couch, 31 March; Wells to Couch, 9 June; Tomlinson to Couch, 7 July. Garden Club Archives, 1932, Minutes of the Executive Board, 15 September. Garden Club of North Carolina, 1931, p. 18; 1932, p. 3. Totten, Ballard, and Little, 1959, p. 4.

14. UNC Press Records, 1932, Couch to Tomlinson, 13 September, 28 September, 4 October; Tomlinson to Couch, 14 September; Iden to Couch, 16 September; Couch to Wells, 4 October.

15. UNC Press Records, 1932, Couch to Tomlinson, 7 October, 5 November, 3 December; Wells to Couch, 25 October; Couch to Mrs. J. G. Walker, 19 November; Couch to L. H. Jenkins, 25 November; Jenkins to Couch, 28 November, 1 December.

16. UNC Press Records, 1931, Tomlinson to Couch, 15 May; 1932, Couch to Wells, 3 December, 9 December; Wells to Couch, 6 December; Tomlinson to Couch, 8 December; Seeman Printery to Couch, 14 December.

17. UNC Press Records, 1932, Couch to Tomlinson, 4 October, 7 October, 5 November; Tomlinson to Couch, 5 October, 7 November; 1933, Tomlinson to Couch, 14 January, 8 March; Couch to Tomlinson, 18 January. Garden Club Archives, 1933, Minutes of the Executive Board, January.

18. UNC Press Records, 1932, Iden to Couch, 13 December; Tomlinson to Couch, 14 December; Couch to Tomlinson, 15 December; Wells to Couch, 21 December, 26 December; Couch to Iden, 31 December.

19. UNC Press Records, 1932, Couch to Jenkins, 25 November; 1933, Jenkins to Couch, 7 March, 21 March; Couch to Jenkins, 13 March; 1934, Jenkins to Couch, 3 July, 6 July; Couch to Jenkins, 5 July. Personal communication, D. B. Anderson, January 1977.

20. UNC Press Records, 1933, Couch to Tomlinson, 13 March; 1934, Jenkins to Couch, 3 July. Garden Club of North Carolina, 1934, p. 35; 1935, p. 43; 1936, p. 14; 1937, p. 20; 1938, p. 15; 1939, p. 14.

21. UNC Press Records, 1933, Couch to Tomlinson, 13 March; 1934, Jenkins to Couch, 3 July.

22. Couch, 1946, p. xxi.

Chapter 8

1. Garden Club of North Carolina, 1933, p. 28; 1935, p. 43. UNC Press Records, 1935, Tomlinson to Elizabeth Lawrence, 15 March; 1942, Couch to Wells, 10 September.

2. UNC Press Records, 1933, Couch to Mrs. J. G. Walker, 10 January; 1934, Couch to R. M. Addoms, 2 July; Couch to H. J. Oosting, 2 July; Addoms to Couch, 8 July; Oosting to Couch, 12 July.

3. UNC Press Records, 1933, Couch to Wells, 2 March; Wells to Couch, 6 March; 1934, Couch to Wells, 9 July; Wells to Couch, 13 July.

4. Wells, A1942. UNC Press Records, 1942, Couch to Wells, 20 September, 21 September; Wells to Couch, 16 September, 29 October.

5. UNC Press Records, 1942, Couch to Wells, 2 November; M. T. Littlejohn to P. L. Ricker, 5 November; 1943, Couch to Wells, 26 March; Wells to Couch, 7 April.

6. *News and Observer*, 1988, William T. Couch, 13 December. UNC Press Records, 1949, Cowles to L. Barber, 2 September; 1951, Cowles to C. Johnson, 21 August; 1959, Davis to Wells, 14 May.

7. UNC Press Records, 1942, P. L. Ricker to UNC Press, 3 November; M. T. Littlejohn to Ricker, 5 November; 1951, C. Johnson to UNC Press, August; 1953, A. Neely to J. H. Benjamin, 9 July; Davis to Neely, 10 July; 1955, E. P. Weeks to UNC Press, 29 November; 1956, O. B. Burwell to Davis, 2 November; Cowles to Burwell, 21 November; Willis Book and Stationery Co. to UNC Press, 3 December. Wells Papers, 1951, S. W. Weaver to Wells, 23 March.

8. Wells Papers and UNC Press Records, 1959, B. Smith to Davis, 12 May; Wells to Davis, 19 May. Wells Papers, 1951, Mrs. H. P. Smith to Wells, 24 August; 1959, B. W. Smith to Wells, 2 December.

9. Wells Papers and UNC Press Records, 1959, Davis to Wells, 14 May, 20 May; Wells to Davis, 19 May.

10. Wells, F1958. UNC Press Records, 1959, Cowles to L. Richardson, 10 November. Sharpe, 1965.

11. UNC Press Records, 1935, Couch to E. D. Fowler, 14 May; Fowler to Couch, 25 May; 1965, Davis to B. Sharpe, 18 February. Personal communication, M. B. Wells, 23 June 1986.

12. UNC Press Records, 1965, Sharpe to Davis, 22 February; Davis to Sharpe, 3 March.

13. Radford, Ahles, and Bell, 1968. Wells Papers and UNC Press Records, 1965, Davis to Wells, 26 March.

14. Wells Papers and UNC Press Records, 1965, Davis to Wells, 26 March, 5 April; Wells to Davis, 28 March, 6 April.

15. UNC Press Records, 1965, Davis to D. B. Anderson, 16 April, 23 April; Anderson to Davis, 21 April; 1967, Page proofs of *The Natural Gardens of North Carolina*, 6 October.

16. Radford, Adams, Bell, Greulach, and Olive, 1973. Estes, 1986. Personal communication, M. B. Wells, 23 June 1986.

17. North Carolina Wild Flower Preservation Society, 1961, 5(1), p. 2; 1967, 8(2), p. 3; 1967, 8(3), pp. 4–5. Personal communication, M. B. Wells, 23 June 1986.

18. UNC Press Records, 1967, Wells to Cowles, 3 September; Davis to Wells, 10 October; Wells to Davis, 13 October.

19. UNC Press Records, 1932, Wells to Couch, 25 October; Couch to Jenkins, 3 December; 1967, M. Boyd to Wells, 2 November. North Carolina Wild Flower Preservation Society, 1967, 8(3)(extra), pp. 3–4; 1968, 8(4), p. 10. Personal communication, M. B. Wells, 23 June 1986.

20. W., 1967. Anonymous, 1968. Lawrence, 1968. Core, 1968.

21. UNC Press Records, 1933, V. N. Walker to Couch, 10 January, 27 January; Couch to Walker, 11 January; Seeman to Couch, 13 January; Couch to M. W. Henry, 10 October; Job P. Wyatt & Sons Co. to Couch, 7 December; 1936, L. K. Miller to UNC Press, 18 June; 1937, R. B. Etheridge to Couch, 12 February. Deans, 1936. deLaubenfels, 1949, p. 157.

22. Wells Papers, 1971, H. J. Rogers to J. Smith, 3 March; Smith to Wells, 9 March. Davis, 1975; 1976. University of North Carolina at Greensboro, 1976. Personal communication, H. J. Rogers, 3 April 1986.

23. Personal communications, L. M. Dellwo, 27 June 1986; R. Maner, 29 May 1992.

Chapter 9

1. O'Keef, 1955. Whitford Papers, Notes.

2. Whitford Papers, Notes. Personal communication, D. B. Anderson, 11 February 1978.

3. Whitford Papers, Notes. Wells, D1952.

4. *North Carolina State College Catalog*, 1919–20, pp. 6–11. North Carolina Library Commission, 1921, p. 24. Lockmiller, 1939, p. 286. Gatewood, 1960b, pp. 227–28. Personal communication, D. B. Anderson, 11 February 1978.

5. Ford, McVey, and Works, 1932, p. 53.

6. *Technician*, 1933, Profs organize State chapter of association, 13 January. Lockmiller, 1939, p. 136. American Association of University Professors, 1932. Carpenter and Colvard, 1987, p. 171.

7. Whitford Papers, Notes. Martin, 1924. Martin and Barkley, 1961. Personal communication, D. B. Anderson, 11 February 1978.

8. Whitford Papers, Notes. Ohio State Academy of Science, 1921, 7(6), p. 185. Wells Papers, no date, Biographical Data.

9. Whitford Papers, Notes. Personal communication, D. B. Anderson, 11 February 1978.

10. Haldane, 1923. Whitford Papers, Notes.

11. Robertson, 1924. Lockmiller, 1939, p. 157. Gatewood, 1960b, pp. 230, 234, 242–43; 1966, p. 16. Thompson, 1932. Anonymous, 1932. Reagan, 1987, pp. 67, 70–71, 88–90. Carpenter and Colvard, 1987, pp. 199, 244. Whitford Papers, Notes; Letters, 1962, L. A. Whitford to H. C. Brown, 29 October.

12. Whitford Papers, Notes. Personal communications, H. T. Scofield, 22 January 1978; M. B. Wells, 22 September 1979, September 1981, 8 December 1984.

13. Lockmiller, 1942, pp. 11–12. Gatewood, 1960b, pp. 254–57. Morrison, 1971, pp. 27, 45. Reagan, 1987, pp. 91–100. Carpenter and Colvard, 1987, pp. 240–41. Whitford Papers, Notes. Personal communications, L. A. Whitford, 2 July 1979; M. B. Wells, September 1981, 8 December 1984.

14. Lockmiller, 1942, p. 72. Ashby, 1980, p. 115. Whitford Papers, Letters, 1962, C. H. Bostian to H. C. Brown, 25 October. UNC Press Records, 1934, Wells to Couch, 13 July. NCSU Archives, 1934, Report of the Curriculum Committee, 14 April. Personal communications, D. B. Anderson, January 1977, 11 February 1978.

15. Personal communication, D. B. Anderson, 11 February 1978.

16. Harrelson, 1944. Whitford Papers, Notes. Personal communications, D. B. Anderson, 11 February 1978; H. T. Scofield, 22 March 1979; L. A. Whitford, 20 July 1978.

17. *News and Observer*, 1925, Additional roster of salaries paid by state, 26 February. Gatewood, 1960b, p. 231. Wilson, 1964, pp. 124–27. Files of the College of Agriculture and Life Sciences, North Carolina State University, 1946–54, Personnel file of B. W. Wells.

18. Harrelson, 1944. Craven and Cate, 1955, p. 563. Reagan, 1987, p. 116. UNC Press Records, 1943, Wells to Couch, 7 April. Personal communications, D. B. Anderson, January 1977, 11 February 1978, 18 April 1982.

19. Craven and Cate, 1955, p. xxxiii. Personal communications, D. B. Anderson, January 1977, 11 February 1978, 18 April 1982.

20. Whitford Papers, Notes. Committee on the Teaching of Botany in American Colleges and Universities, 1938, p. 26.

21. Whitford Papers, Notes. Personal communications, D. B. Anderson, 11 February 1978, 18 April 1982; T. L. Quay, 24 March 1992; M. B. Wells, 20 May 1992.

22. *News and Observer*, 1951, Death claims Doctor Shunk, 12 July. Wells, D1952. Lynch, 1952. Godfrey, 1953. Whitford Papers, Notes. Personal communications, D. B. Anderson, 11 February 1978, 18 April 1982; H. T. Scofield, 22 January 1978.

Chapter 10

1. Whitford Papers, Notes. Wells Papers, 1978, audiotape, T. L. Quay.

2. Lockmiller, 1939, p. 286. *North Carolina State College Catalog*, 1919–20, pp. 106–8.

3. Whitford Papers, Notes. *North Carolina State College Catalog*, 1931–32, p. 185; 1932–33, p. 103.

4. *North Carolina State College Catalog*, 1931–32, p. 186; 1932–33, p. 164; 1934–35, p. 158; 1945–46, p. 187.

5. *North Carolina State College Catalog*, 1919–20 through 1953–54.

6. Wells, F1923b. *State College Record*, 1936, Forestry announcement 1937–38, 36(2), pp. 5–7. Whitford Papers, Letters, 1962, S. G. Boyce to H. C. Brown,

20 November. Files of the Department of Botany, North Carolina State University, 1978, A. B. Crowe to Department of Botany, 6 January. Stick, 1985, p. 98.

7. Whitford Papers, Notes. Lockmiller, 1939, p. 287. Wells Papers, 1978, audiotape, T. L. Quay. Shelley, 1974. Whitford Papers, Letters, 1962, S. G. Boyce to H. C. Brown, 20 November; L. A. Whitford to H. C. Brown, 12 November.

8. Heltzel, 1936. Davis, 1940. West, 1948. Renfro, 1951. Reid, 1952. Raper, 1954.

9. NCSU Archives, 1924–53, Commencement programs. Williams, 1924. Martin, 1924. Whitford, 1929. Roller, 1931. Crafton, 1932. Godfrey, 1938. McMenamin, 1940. Cappell, 1953. Boyce, 1953. Whitford Papers, Letters, 1963, G. F. Papenfuss to H. C. Brown, 28 January.

10. Personal communication, L. A. Whitford, 16 February 1978. Wells Papers, 1978, audiotape, S. G. Boyce, T. L. Quay. Whitford Papers, Letters, 1962, R. K. Godfrey to H. C. Brown, 13 November; 1963, G. F. Papenfuss to H. C. Brown, 28 January.

11. Whitford Papers, Notes. *Technician*, 1933, Dr. Wells talks to Ag freshmen on study means, 27 October. North Carolina Academy of Science, 1924, p. 97. Mountain Lake Biological Station, 1939; 1950. Hardin, 1984, p. 5.

12. Anonymous, 1931a; 1931b; 1931c; 1936; 1938a; 1938b; 1938d. Bagley, 1935, pp. 270–71. Robson, 1937, p. 138. Wells, E1935. Watson, 1964. Wells Papers, 1965, New College Directory, T. Alexander, p. 1.

13. Watson, 1964. Personal communication, L. A. Whitford, 14 June 1979. LaFollette, 1935.

14. Personal communications, M. B. Wells, 22 September 1979, 23 June 1986. Anonymous, 1933b. LaFollette, 1935. *Technician*, 1933, Botany head to teach summer school course, 26 May. Wells, E1935.

15. Wells, E1935. Wells Papers, 1965, New College Directory, B. W. Wells, p. 10.

16. Wells, E1918. Wells, Shunk, and Martin E1922[?]. Whitford Papers, L. A. Whitford, 1921–22, Laboratory notebook in botany; 1939–40, Laboratory directions and review questions in general botany.

17. Wells, E1921. Files of the Department of Botany, North Carolina State University, 1980, F. B. Monroe to Department of Botany, 5 July.

18. Wells, E1921. Wells Papers, 1978, audiotape, J. M. Clarkson. Shelley, 1974. Whitford Papers, Letters, 1962, A. W. Bayer to H. C. Brown, 22 November. O'Keef, 1955.

Chapter 11

1. Vanderlaan, 1925. Shipley, 1927. Furniss, 1963. Gatewood, 1966.

2. Knight, 1927. Shipley, 1927. Wilson, 1957; 1960. Gatewood, 1966. Linder, 1963; 1966. Troyer, 1987.

3. Gatewood, 1966, pp. 48–51. Black, Coker, and Bradbury, 1938. *News and Observer*, 1922, 7–13 May.

4. *News and Observer*, 1922, Reply to Massee about evolution, 14 May. Gatewood, 1966, pp. 51–52. Reagan, 1987, pp. 86–87.

5. *News and Observer*, 1922, Riley prepared to meet professors, 15 May; Debate evolution at Pullen Hall tomorrow afternoon, 16 May. Gatewood, 1966, pp. 52–53.

6. *News and Observer*, 1922, Debate evolution at Pullen Hall tomorrow afternoon, 16 May; Scientist and theologian to debate evolution today, 17 May. Whitford Papers, Notes.

7. *News and Observer*, 1922, Vociferous demonstrations accompany evolution debate, 18 May. Gatewood, 1969, pp. 157–61.

8. *News and Observer*, 1922, Vociferous demonstrations accompany evolution debate, 18 May. Gatewood, 1969, p. 159.

9. *News and Observer*, 1922, Vociferous demonstrations accompany evolution debate, 18 May. Gatewood, 1969, pp. 160–61.

10. *News and Observer*, 1922, Vociferous demonstrations accompany evolution debate, 18 May. Gatewood, 1969, pp. 159–60.

11. *News and Observer*, 1922, Vociferous demonstrations accompany evolution debate, 18 May. Gatewood, 1969, p. 161.

12. *News and Observer*, 1922, Vociferous demonstrations accompany evolution debate, 18 May. Gatewood, 1969, pp. 157–61. Whitford Papers, Undergraduate Years.

13. *News and Observer*, 1924, 24 January. Gatewood, 1966, p. 106.

14. North Carolina Academy of Science, 1924. Troyer, 1987.

15. *News and Observer*, 1925, Bill outlawing Darwinism fails by only one vote, 11 February. Wells and Metcalf, D1925. Gatewood, 1966, pp. 124–32.

16. *News and Observer*, 1925, Bill outlawing Darwinism fails by only one vote, 11 February. *Technician*, 1925, Anti-Darwinism bill reported unfavorably, 13 February. Wells and Metcalf, D1925. Gatewood, 1966, pp. 133–34. Linder, 1966, pp. 130–33.

17. Lewis, 1926, Section X. Wells and Metcalf, D1925. Gatewood, 1966, p. 139.

18. Knight, 1927. Shipley, 1927, pp. 93–95. Gatewood, 1966, pp. 157–58.

19. North Carolina Academy of Science, 1926, pp. 6–7. Shipley, 1927, pp. 99–100. Troyer, 1987.

20. Wells, D1926. Shipley, 1927, p. 109.

21. *News and Observer*, 1927, 26 January, 11 February, 24 February. Gatewood, 1966, pp. 221–29. NCSU Archives, 1927, Minutes of the Faculty Council, 1 February.

Chapter 12

1. O'Keef, 1955. Wells Papers, no date, Record of B. W. Wells. Whitford Papers, Notes. Personal communications, M. B. Wells, 22 September 1979; L. A. Whitford, 14 June 1979.

2. Whitford Papers, Notes. Personal communications, M. B. Wells, 22 September 1979; L. A. Whitford, 2 July 1979.

3. Wells Papers, no date, Record of B. W. Wells. *Technician*, 1925, Facultyfax, Dr. Bertram Whittier Wells, 27 March. Wells, F1932/1967, p. 326. O'Keef, 1955.

4. Halstead, 1901, pp. 245–48. Everett, 1901, pp. 413–23. Personal communication, M. B. Wells, 22 September 1979.

5. Wells Papers, no date, Record of B. W. Wells. Personal communication, M. B. Wells, 22 September 1979.

6. Whitford Papers, Notes. Personal communications, M. B. Wells, 22 September 1979, 3 March 1981; L. A. Whitford, 16 February 1978.

7. Personal communications, D. B. Anderson, 12 February 1981; M. B. Wells, 22 September 1979; L. A. Whitford, 14 June 1979. *News and Observer*, 1925, 11 February. Wells Papers, 1978, audiotape, T. L. Quay.

8. McCray, 1909. Wells, D1941b. Griggs, 1941.

9. McCray, 1909. Wells, D1911a; D1911b; D1911c; D1912. Dickey, 1911.

10. Dickey, 1910. Griggs, 1961, pp. 50–51. Meyer, 1983, p. 41. Gurney, 1961.

11. Wells Papers, no date, Record of B. W. Wells. Ohio State Academy of Science, 1911, 6(1), p. 10; 1912, 6(2), p. 53; 1913, 6(3), p. 99; 1914, 6(4), p. 153. Personal communication, M. B. Wells, 22 September 1979.

12. O'Keef, 1955. Personal communication, D. B. Anderson, 18 April 1982.

13. Wells Papers, no date, Record of B. W. Wells. Wells, B1914; B1915; B1916. Ohio State Academy of Science, 1911, 6(1), p. 10; 1915, 6(5), p. 216; 1930, 8(7), p. 349. Alexander, 1941, p. 304.

14. Wells Papers, no date, Record of B. W. Wells. Ohio State Academy of Science, 1916, 7(1), p. 12. Personal communication, M. B. Wells, 22 September 1979.

15. Ohio State Academy of Science, 1917, 7(2), p. 37; 1918, 7(3), p. 70. Wells, C1919. Ousley, 1935, p. 56. *The College Blue Book*, 1977, p. 710. Personal communication, M. B. Wells, 22 September 1979.

16. Wells Papers, no date, Record of B. W. Wells. Anonymous, 1921. Hale, 1948, pp. 10, 101, 214.

Chapter 13

1. Whitford Papers, Notes. Personal communications, D. B. Anderson, 11 February 1978; H. T. Scofield, 22 January 1978; M. B. Wells, 22 September 1979. Flory, 1987, p. 13.

2. Wells Papers, no date, Record of B. W. Wells. Personal communications, D. B. Anderson, 11 February 1978; M. B. Wells, 22 September 1979, 24 July 1981, 31 July 1981, 20 May 1992, 23 June 1992.

3. Nygard, 1989. Whitford Papers, Notes. Painter, 1977. Wells Papers, 1978, audiotape, A. W. Cooper, T. L. Quay. Cooper, 1979a; 1979b.

4. Whitford Papers, Notes. Gurney, 1961. Felt, 1940, pp. 16, 17, 129, 232. Dachnowski, 1912. Personal communication, L. A. Whitford, 2 July 1979. Wells Papers, no date, Record of B. W. Wells.

5. *Technician*, 1936, Women's Club to hear Wells at art exhibit, 7 February. Shelley, 1974. Personal communications, S. G. Boyce, September 1981; M. B. Wells, 8 December 1984. Wells Papers, no date, Record of B. W. Wells.

6. NCSU Archives, 1934, Report of the Curriculum Committee, 14 April.

7. NCSU Archives, 1934, Minutes of the Library Committee. Kirk, Cutter, and Morse, 1936, p. 253. Whitford Papers, Notes. Jones, 1989.

8. *News and Observer*, 1934, Art exhibit opens Monday, 2 December. Wells,

F1934. *Technician*, 1934, Library murals will be placed in coming week, 7 December.

9. Denmark, 1934. *Technician*, 1935, Our modernistic murals, 11 January; McLean murals elicit comments from teachers, Murals bring varied reactions, 25 January; Faculty to pass decision on disputed decorations before final placement, 1 February. Vess, 1982a.

10. *News and Observer*, 1935, Our modernistic murals, 21 January. Henderson, 1935. Lockmiller, 1939, p. 203. Federal Writers' Project, 1939, p. 246.

11. NCSU Archives, 1941, Minutes of the Library Committee, 7 February; Minutes of the Faculty, 12 September. *News and Observer*, 1941, Murals stored to await new building, 23 September. Jones, 1989. Marshall, 1941.

12. *News and Observer*, 1982, '30s mural, once cast aside, returns to N.C. State, 3 October; Back again, 14 October. Vess, 1982a; 1982b. *Technician*, 1982, McLean dedication least State could do, 11 October. Anonymous, 1982. *Journal*, Artist's work returned to university, 14(3), p. 3. Litt, 1985.

13. Connor, 1938. *Technician*, 1938, Professor's wife taken by death, 11 February. Cunningham, Anderson, and Metcalf, 1938. Wells, C1921. Personal communications, D. B. Anderson, 11 February 1978, 18 April 1982.

14. Whitford Papers, Notes. Personal communications, D. B. Anderson, 18 April 1982; L. A. Whitford, 14 June 1979.

15. Whitford Papers, Notes. Personal communications, M. B. Wells, 22 September 1979; L. A. Whitford, 16 February 1978. Shelley, 1974.

16. Personal communication, M. B. Wells, 22 September 1979. Wells Papers, 1978, audiotape, K. M. Browne.

17. Shelley, 1974. Personal communications, D. B. Anderson, January 1977, 18 April 1982; M. B. Wells, 23 June 1986; L. A. Whitford, 14 June 1979. *News and Observer*, 1991, W. L. Wells, early surgeon at UNC-CH, 29 May. *American Men of Science*, 1966, p. 2594 (Frank Johnson). Wells Papers, 1981, F. H. Johnson to M. B. Wells, 16 March.

18. Whitford Papers, Notes. Wells Papers, no date, Record of B. W. Wells. Personal communications, D. B. Anderson, 18 April 1982; M. B. Wells, 22 September 1979.

19. Whitford Papers, Notes. Personal communications, D. B. Anderson, 11 February 1978; H. T. Scofield, 22 January 1978; M. B. Wells, 22 September 1979, 20 May 1992. Wells Papers, 1978, audiotape, J. M. Clarkson; 1978, D. B. Anderson to M. B. Wells, 31 December.

20. Whitford Papers, Notes. Personal communications, D. B. Anderson, 11 February 1978; M. B. Wells, 22 September 1979.

21. Connor, 1938. *Technician*, 1938, Professor's wife taken by death, 11 February. Cunningham, Anderson, and Metcalf, 1938. Personal communications, D. B. Anderson, 11 February 1978, 18 April 1982; M. B. Wells, 22 September 1979, 20 May 1992. Brown, 1978.

22. Whitford Papers, Notes. Gatewood, 1960b, p. 255. NCSU Archives, 1932, Commencement program. *News and Observer*, 1941, Miss Maude Rhodes Barnes married to Dr. B. W. Wells, 26 February. Personal communications, M. B. Wells, 22 September 1979, 24 July 1981, 23 June 1986, 20 May 1992.

23. Personal communications, M. B. Wells, 22 September 1979, 20 May 1992.

24. *News and Observer*, 1941, Miss Maude Rhodes Barnes married to Dr. B. W. Wells, 26 February; Officers of state council, 20 January. Personal communication, M. B. Wells, 22 September 1979. Whitford Papers, Notes. Wells, A1946a. Cooper, 1979a.

25. Personal communications, D. B. Anderson, 11 February 1978, 18 April 1982; A. W. Cooper, August 1984; H. T. Scofield, 22 January 1978; B. W. Wells, April 1957; L. A. Whitford, 16 February 1978. Wells, A1949. Wells and Boyce, A1953c, p. 121. Wells Papers, 1971, B. W. Wells to R. Jastrow, 16 July. Whitford Papers, Notes. Personal experience of the author.

26. Personal communications, D. B. Anderson, 11 February 1978; H. T. Scofield, 22 January 1978; L. A. Whitford, 16 February 1978; R. L. Wilbur, April 1957.

Chapter 14

1. Wells, F1961a. Personal communications, M. B. Wells, 22 September 1979, 2 July 1981.

2. Personal communications, M. B. Wells, 22 September 1979, 20 May 1992. Painter, 1977.

3. Payne, 1977. Nygard, 1989. Personal communications, M. B. Wells, 22 September 1979, 20 May 1992.

4. Wells Papers, 1971, audiotape, B. W. Wells. Payne, 1977. Painter, 1977. Personal communication, M. B. Wells, 22 September 1979.

5. Payne, 1977. Painter, 1977. Nygard, 1989. Personal communication, M. B. Wells, 22 September 1979, September 1981. Wells Papers, 1971, audiotape, B. W. Wells.

6. Shelley, 1974. Payne, 1977. Personal communication, M. B. Wells, 22 September 1979. Wells Papers, 1973, Pictures given to friends; audiotape, 1971, B. W. Wells.

7. Personal communication, M. B. Wells, 22 September 1979. Payne, 1977. Wells Papers, 1971, audiotape, B. W. Wells; no date, Record of B. W. Wells.

8. Wells Papers, no date, The Rockcliff Farm—A Letter to Children; 1962, W. O. Douglas to B. W. Wells, 2 July, 9 July. Personal communication, M. B. Wells, 22 September 1979. Painter, 1977. Payne, 1977.

9. Personal communication, M. B. Wells, 22 September 1979. Wells Papers, 1963, P. S. Warren to B. W. and M. B. Wells, 27 May; no date, The Rockcliff Farm—A Letter to Children.

10. Wells Papers, 1978, audiotape, A. W. Cooper. Whitford Papers, 1968, B. W. Wells to L. A. Whitford, 5 September. Personal communication, M. B. Wells, 22 September 1979. *News and Observer*, 1978, Former professor dies at 94, 30 December; 1979, Funeral held for scientist, 3 January. *Raleigh Times*, 1979, Service held for scientist, 3 January.

11. Cooper, 1979a; 1979b. Troyer, 1986. Garden Club Archives, 1982, Mrs. W. C. Landolina to Mrs. Ruth Reichard, 15 June. Totten, Ballard, and Little, 1959, pp. 21, 77, 78. Wells Papers, 1954, D. W. Colvard to B. W. Wells, 13 August; 1955, H. A. Royster to B. W. Wells, 31 January. O'Keef, 1955.

12. Wells Papers, 1963, J. T. Caldwell to B. W. Wells, 27 February. NCSU Archives, 1963, Commencement program. *Raleigh Times*, 1963, NCS honors

4 with degrees, 1 June. *News and Observer*, 1963, Sanford, 1,222 get degrees at State, 2 June. Reagan, 1987, pp. 58, 183.

13. Shelley, 1974. *Journal*, 1977, Five Botany Dept. heads spanning 58 years honored at banquet, November. Anonymous, 1977; 1978.

14. NCSU Archives, 1979, Minutes of the Board of Trustees, 18 May. North Carolina Wildlife Federation, 1979. *Journal*, 1980, Gardner addition named for Bostian, June; 1981, Auditoriums named for professors, April. North Carolina Agricultural Research Service, 1984. Kirkman and Ballington, 1985. North Carolina State University Student Center, 1987. Earley, 1989, pp. 21, 22.

15. Wells Papers, 1979, D. H. Martin to H. N. Lee, 23 January; J. Lawrence and J. Moore to A. A. Hight, 27 February. B. W. Wells Association, 1983. *Raleigh Times*, 1982, Charters, 25 May. Parkins, 1982. Simpson, 1982. *News and Observer*, 1986, Wells Association wins parks award, 21 April. Green, 1986. Nygard, 1989.

BIBLIOGRAPHY

I. Unpublished Sources

Garden Club of North Carolina Archives. Cited as "Garden Club Archives," these records are held by the Garden Club of North Carolina, Inc., 505 Oberlin Road, Raleigh, N.C. 27605.

North Carolina State University Archives. Cited as "NCSU Archives," these sources include the report of a curriculum committee, commencement programs, and minutes of the board of trustees, the faculty, the faculty council, and the library committee; they are held in University Archives, D. H. Hill Library, North Carolina State University, Raleigh, N.C. 27695.

Personal Communications. Information supplied by an individual directly to the author, either orally or in writing is cited as "Personal communication."

Records of the University of North Carolina Press. Cited as "UNC Press Records," these documents pertaining to the publication and republication of *The Natural Gardens of North Carolina* are held in Subgroup 4, University Archives, Manuscripts Department, University of North Carolina Library, Chapel Hill, N.C. 27514.

Wells Papers. The few remaining papers of B. W. Wells, in the possession of and made available by Maude Barnes Wells, consist of letters, biographical summaries, and miscellaneous documents; also included are two audiotapes, one made in 1971 by Wells himself, and one made after his death by a group of his friends on 31 December 1978.

Whitford Papers. The personal papers made available by Larry A. Whitford consist of various documents, but especially important are "Notes for a History of the Botany Department, North Carolina State University at Raleigh," unpublished manuscript, 1970, cited as "Notes"; "My Undergraduate Years at North Carolina State College," unpublished manuscript, 1984, cited as "Undergraduate Years"; and copies of letters written in support of the candidacy of B. W. Wells for an honorary degree from North Carolina State College, cited as "Letters."

Other unpublished sources are identified completely in the notes.

II. Published Writings of B. W. Wells

A. Reports of Research in Ecology

A1923. Savannah and sand ridge plant communities. *Journal of the Elisha Mitchell Scientific Society* 39: 14–15. (Abstract.)

A1924a. The ecological position of the eastern North Carolina pine communities. *Journal of the Elisha Mitchell Scientific Society* 40: 103–4. (Abstract.)

A1924b. *Major Plant Communities of North Carolina*. North Carolina Agricultural Experiment Station, Technical Bulletin No. 25. 20 pp.

A1926. A southern upland bog succession. *Journal of the Elisha Mitchell Scientific Society* 42: 9. (Title only.)

A1928a. Plant communities of the coastal plain of North Carolina and their successional relations. *Ecology* 9: 230–42.

A1928b. B. W. Wells and I. V. D. Shunk. *A Southern Upland Grass-Sedge Bog: An Ecological Study*. North Carolina Agricultural Experiment Station, Technical Bulletin No. 32. 75 pp.

A1929a. The "open grounds" of Carteret Co., N.C. *Journal of the Elisha Mitchell Scientific Society* 45: 19–20. (Abstract.)

A1929b. A. P. Dachnowski-Stokes and B. W. Wells. The vegetation, stratigraphy, and age of the "open land" peat area in Carteret County, North Carolina. *Journal of the Washington Academy of Sciences* 19: 1–11.

A1931. B. W. Wells and I. V. D. Shunk. The vegetation and habitat factors of the coarser sands of the North Carolina coastal plain: an ecological study. *Ecological Monographs* 1: 465–520.

A1932. Soil water relations in the sand hills. *Journal of the Elisha Mitchell Scientific Society* 47: 11. (Title only.)

A1934. W. M. Crafton and B. W. Wells. The old field prisere: an ecological study. *Journal of the Elisha Mitchell Scientific Society* 49: 225–46 + 3 plates.

A1936a. Origin of the southern Appalachian grass balds. *Bulletin of the Ecological Society of America* 17: 23. (Abstract.)

A1936b. Origin of the southern Appalachian grass balds. *Science* 83: 283.

A1936c. Andrews Bald: the problem of its origin. *Journal of the Southern Appalachian Botanical Club* 1: 59–62.

A1937a. Southern Appalachian grass balds. *Journal of the Elisha Mitchell Scientific Society* 53: 1–26 + 5 plates.

A1937b. B. W. Wells and I. V. D. Shunk. Seaside shrubs: wind forms vs. spray forms. *Science* 85: 499.

A1938a. Southern Appalachian grass balds as evidence of Indian occupation. *Bulletin of the Archeological Society of North Carolina* 5: 2–7.

A1938b. B. W. Wells and I. V. [D.] Shunk. The important role of salt spray in coastal ecology. *Journal of the Elisha Mitchell Scientific Society* 54: 185–86. (Abstract.)

A1938c. B. W. Wells and I. V. D. Shunk. Salt spray: an important factor in coastal ecology. *Bulletin of the Torrey Botanical Club* 65: 485–92.

A1938d. A new forest sub-climax: the salt spray climax of Smith Island, North Carolina. *Bulletin of the Ecological Society of America* 19: 35–36. (Abstract.)

A1939. A new forest climax: the salt-spray climax of Smith Island, N.C. *Bulletin of the Torrey Botanical Club* 66: 629–34.

A1940. Preliminary survey of the eastern Dare County peat. *Journal of the Elisha Mitchell Scientific Society* 56: 219–20. (Abstract.)

A1942. Ecological problems of the southeastern United States coastal plain. *Botanical Review* 8: 533–61.

A1946a. *Vegetation of Holly Shelter Wildlife Management Area.* North Carolina Department of Conservation and Development, Division of Game and Inland Fisheries, State Bulletin No. 2. 40 pp.
A1946b. Archeological disclimaxes. *Journal of the Elisha Mitchell Scientific Society* 62: 51–53.
A1949. Origin of the Carolina Bays: evidence from peat profiles. *Journal of the Elisha Mitchell Scientific Society* 65: 185. (Abstract.)
A1952. Phytogeographical derivation of shrub-bog and swamp forest species. *Journal of the Elisha Mitchell Scientific Society* 68: 132. (Title only.)
A1953a. B. W. Wells and S. G. Boyce. Why the "whiteness" of White Lake, Bladen Co., N.C. *Journal of the Elisha Mitchell Scientific Society* 69: 90. (Abstract.)
A1953b. B. W. Wells and S. G. Boyce. The Bladen Lakes (N.C.): advancing or retreating bog margins? *Journal of the Tennessee Academy of Science* 28: 190. (Abstract.)
A1953c. B. W. Wells and S. G. Boyce. Carolina Bays: additional data on their origin, age, and history. *Journal of the Elisha Mitchell Scientific Society* 69: 119–41.
A1954. B. W. Wells and S. G. Boyce. Carolina Bay lakes: the bog margin problem. *Ecology* 35: 584.
A1956. Origin of southern Appalachian grass balds. *Ecology* 37: 592.
A1961. The southern Appalachian grass bald problem. *Castanea* 26: 98–100.
A1976[1977]. B. W. Wells and L. A. Whitford. History of stream-head swamp forests, pocosins, and savannahs in the Southeast. *Journal of the Elisha Mitchell Scientific Society* 92: 148–50.

B. Reports of Research on Insect Galls

B1914. Some unreported cecidia from Connecticut. *Ohio Naturalist* 14: 289–98.
B1915. A survey of the zoocecidia on species of *Hicoria* caused by parasites belonging to the Eriophyidae and the Itonididae (Cecidomyiidae). *Ohio Journal of Science* 16: 37–59.
B1916. The comparative morphology of the zoocecidia of *Celtis occidentalis*. *Ohio Journal of Science* 16: 249–98.
B1918. The zoocecidia of northeastern United States and eastern Canada. *Botanical Gazette* 65: 535–42.
B1920. Early stages in the development of certain *Pachypsylla* galls on *Celtis*. *American Journal of Botany* 7: 275–85.
B1921a. New North Carolina gall types. *Journal of the Elisha Mitchell Scientific Society* 37: 16. (Title only.)
B1921b. New United States zoocecidia. *Psyche* 28: 35–45 + 1 plate.
B1921c. Evolution of zoocecidia. *Botanical Gazette* 71: 358–77 + 2 plates.
B1921d. B. W. Wells and Z. P. Metcalf. A new species of oak gall and its maker. *Canadian Entomologist* 53: 212–13.
B1921e. Gall evolution: a new interpretation. *Science* 54: 301.
B1923. Fundamental classification of galls. *Science* 57: 469–70.
B1934. Galls and "galls." *Journal of the Elisha Mitchell Scientific Society* 50: 65–74.

C. Miscellaneous Scientific Reports

C1910. A histological study of the self-dividing laminae of certain kelps. *Ohio Naturalist* 11: 217–32.

C1919. [On an aberrant inflorescence of *Allium mutabile*.] *Plant World* 22: 251–52.

C1921. A phenomenal shoot. *Science* 54: 13–14.

C1928. A new species of *Pyxidanthera*. *Journal of the Elisha Mitchell Scientific Society* 44: 11. (Title only.)

C1929. A new pyxie from North Carolina. *Journal of the Elisha Mitchell Scientific Society* 44: 238–39 + 1 plate.

C1941. B. W. Wells and I. V. D. Shunk. The organic deposits of the lower Cape Fear peninsula. *Journal of the Elisha Mitchell Scientific Society* 57: 197–98. (Abstract.)

C1943. Blythe Bay: a record of changing ocean levels. *Journal of the Elisha Mitchell Scientific Society* 59: 118–19. (Abstract.)

C1944. Origin and development of the lower Cape Fear peninsula. *Journal of the Elisha Mitchell Scientific Society* 60: 129–34 + 3 plates.

C1947. A bay in a bay. *Journal of the Elisha Mitchell Scientific Society* 63: 93. (Title only.)

D. Other Communications to Scientists

D1911a. Meetings of the Biological Club. November 7, 1910. *Ohio Naturalist* 11: 287–88.

D1911b. Meetings of the Biological Club. January 16, 1911. *Ohio Naturalist* 11: 352.

D1911c. Meetings of the Biological Club. March 7, April 4, 1911. *Ohio Naturalist* 11: 384.

D1912. Meetings of the Biological Club. May 1, 1911. *Ohio Naturalist* 12: 458.

D1925. B. W. Wells and Z. P. Metcalf. Teaching of evolution in North Carolina. *Science* 61: 445.

D1926. Fundamentalism in North Carolina. *Science* 64: 17–18.

D1927. College research. *North Carolina Experiment Station Annual Report*: 105–7.

D1938a. Easter field trip to Wilmington, North Carolina. April 16–18, 1938. *Torreya* 38: 21–22.

D1938b. Southern Appalachian and Torrey Botanical Club Easter foray in Wilmington region. *Castanea* 3: 73–77.

D1941a. Plant galls. Review of E. P. Felt, *Plant Galls and Gall Makers*. *Ohio Journal of Science* 41: 250.

D1941b. [Tribute to J. H. Schaffner.] In Professor John Henry Schaffner, by A. E. Waller. *Ohio Journal of Science* 41: 253–86.

D1952. Ivan Vaughan Detweiler Shunk. *Journal of the Elisha Mitchell Scientific Society* 68: 128–29.

D1953. The Southern Appalachian Botanical Club in 1953. III. Foray into the North Carolina Coastal Plain, April 18–19, 1953. *Asa Gray Bulletin* 2: 201–7.

E. Works Related to Teaching

E1918. *Botany Laboratory Guide for Elementary and General Botany Courses*. Fayetteville, Ark.: The author. 40 pp.

E1921. A method of teaching the evolution of the land plants. *Torreya* 21: 45–47 + 1 plate.

E1922[?]. B. W. Wells, I. V. D. Shunk, and A. C. Martin. *Laboratory Directions for General Botany*. Raleigh: North Carolina State College of Agriculture and Engineering. 29 pp.

E1935. An adventure in natural science education. *Journal of the Elisha Mitchell Scientific Society* 51: 213–14. (Abstract.)

F. Popular or Semipopular Works

F1923a. A new tea plant for North Carolina. *North Carolina Agriculture and Industry*, 17 October.

F1923b. A forestry school in North Carolina. *North Carolina Agriculture and Industry*, 29 November.

F1924–25. The patchwork of North Carolina's great green quilt. *North Carolina Agriculture and Industry*, 1924, 23 October; 6, 27 November; 11 December; 1925, 22 January; 12 February.

F1926. Z. P. Metcalf and B. W. Wells. North Carolina. In *Naturalist's Guide to the Americas*, edited by V. E. Shelford, pp. 413–18. Baltimore: Williams and Wilkins Company.

F1927. [Chapter on southern wildflowers.] In *The Book of Knowledge; The Children's Encyclopedia*, edited by H. Thompson and A. Mee. New York: Grolier Society.

F[n.d.] *The Remarkable Flora of the Great Smoky Mountains* [Cover title: *Flora of the Great Smoky Mountains*.] Asheville: North Carolina National Park Commission.

F1932/1967. *The Natural Gardens of North Carolina; with Keys and Descriptions of the Herbaceous Wild Flowers Found Therein*. Chapel Hill: University of North Carolina Press. Reprinted 1967 with an addendum and elimination of "a few paragraphs from the Preface and Introduction."

F1934. College murals state's outstanding art achievement in 1934. *News and Observer*, 2 December.

F1936/1959. The plant that has made North Carolina famous. Garden Club of North Carolina, *Yearbook No. 6*, pp. 41–42. Reprinted 1959 in *The First Thirty-Four Years, 1925–1959: A Fact Finder of the Garden Club of North Carolina, Inc.*, edited by Mrs. H. R. Totten, L. Ballard, and Mrs. G. W. Little, p. 3. Raleigh: Garden Club of North Carolina, Inc.

F1939/1955. [Unattributed.] Natural setting. In *A Guide to the Old North State*, by Federal Writers' Project, pp. 8–24. Chapel Hill: University of North Carolina Press. Attributed revision 1955, Natural setting. In *The North Carolina Guide*, edited by B. P. Robinson, pp. 26–49. Chapel Hill: University of North Carolina Press.

F1944. The truth about storms along our coast. *State* 12(17): 4–5.

F1949. North Carolina. In *A Traveler's Guide to Roadside Wild Flowers, Shrubs and Trees of the U.S.*, edited by K. S. Taylor, pp. 135–36. New York: Farrar, Straus and Company.

F1950. The flora and fauna of the upper Cape Fear. In *The Story of Fayetteville and the Upper Cape Fear*, edited by J. A. Oates, pp. 114–18. Fayetteville, N.C. 2d edition 1972 (Reprint + Supplement + Index). Fayetteville, N.C.: Fayetteville Women's Club. 3d edition 1981. Raleigh, N.C.: Fayetteville Women's Club.

F1952. The vegetation of the Southport region. *Chat* 16: 28.

F1956. The distribution of wild flowers in relation to plant succession. *North Carolina Wild Flower Preservation Society Newsletter* 2(4): 1–2.

F1958. Gone forever! When the majestic longleaf thrived. [Excerpt from *The Natural Gardens of North Carolina.*] *State* 26(3): 11.

F1960. The balds. *State* 28(9): 5–6.

F1961a. The case of the puzzled scientist. *State* 28(21): 13–14.

F1961b. The flora of one hundred and fifty acres. *North Carolina Wild Flower Preservation Society Newsletter*, April.

F1967. Some plant communities and habitats of the remarkable Wilmington area. *North Carolina Wild Flower Preservation Society Newsletter* 8(3): 12–14.

F1971. Growing North Carolina wild flowers. *North Carolina Wild Flower Preservation Society Newsletter*, October, pp. 11–13.

III. Other Published Sources

Ahles, H. E. 1964. New combinations for some vascular plants of southeastern United States. *Journal of the Elisha Mitchell Scientific Society* 80: 172–73.

Alexander, W. H. 1941. The Ohio Academy of Science (mostly historical). *Ohio Journal of Science* 41: 288–312.

American Association of University Professors. 1932. Membership list. *Bulletin of the American Association of University Professors* 18(1): 66.

American Men of Science. 1921–66.

Anonymous. 1921. Higher education in Arkansas. *School and Society* 14: 376–78.

———. 1931a. A new teachers college. *School and Society* 34: 528.

———. 1931b. A "new" teachers college. *Survey* 67: 208.

———. 1931c. The new college for teachers at Columbia University. *Journal of Home Economics* 23: 1145–46.

———. 1932. The North Carolina State College and Dean Carl C. Taylor. *School and Society* 35: 117–18.

———. 1933a. The Natural Gardens of North Carolina. *Science News Letter* 23: 159.

———. 1933b. Putting college students on farms to fit them for life. *Literary Digest*, 16 September, p. 21.

———. 1936. A new residential secondary school in North Carolina. *School and Society* 44: 473.

———. 1938a. New College of Teachers College, Columbia University. *School and Society* 47: 175.

———. 1938b. The discontinuance of New College, Columbia University. *School and Society* 48: 651.

———. 1938c. Trouble at T. C. *Time*, 28 November, pp. 36–37.

———. 1938d. New College of Teachers College, Columbia University. *School and Society* 48: 761–62.
———. 1951. Cape Fear Club has program on conservation. *North Carolina Gardener* 20(3): 12.
———. 1968. UNC Press publishes revisions of major gardening works. *Durham Morning Herald*, 14 January.
———. 1977. 5 NCSU Botany Department heads honored. *Statelog*, November. Raleigh: North Carolina State University.
———. 1978. Botany staff honors five department heads. *SALS Revue*, Spring. Raleigh: School of Agriculture and Life Sciences, North Carolina State University.
———. 1982. Art of forgiveness returns to campus. *Statelog*, December. Raleigh: North Carolina State University.
Ashby, W. 1980. *Frank Porter Graham: A Southern Liberal*. Winston-Salem, N.C.: John F. Blair Publisher.
B. W. Wells Association. 1983. Brief history. Activities. *Newsletter*, Fall, pp. 1–3.
Bagley, C. T. 1935. The price of poor teaching. *Survey Graphic* 24: 223–26, 270–71.
Baldwin, S. 1968. *Poverty and Politics: The Rise and Decline of the Farm Security Administration*. Chapel Hill: University of North Carolina Press.
Ballington, J. R. 1984. The history of blueberry improvement in North America. In *Proceedings of the Fifth North American Blueberry Research Workers Conference*, edited by T. E. Crocker and P. Lyrene, pp. 8–14. Gainesville: University of Florida.
Barker, L., and V. A. Zullo. 1980. *Asterias forbesi* (Desor) (Asterozoa, Asteroidea) from the Pleistocene "coquina" at Fort Fisher, Hanover County, North Carolina. *Journal of the Elisha Mitchell Scientific Society* 96: 39–44.
Bayer, A. W. 1938. An account of the plant ecology of the coastal belt and midlands of Zululand. *Annals of the Natal Museum* 8: 271–454.
Biblical Recorder. 1932. 21 December. Raleigh, N.C.
Billings, W. D. 1970. Resolution of respect, Henry J. Oosting, 1903–1968. *Bulletin of the Ecological Society of America* 51(14): 16–17.
Black, C. S., W. C. Coker, and O. C. Bradbury. 1938. William Louis Poteat. *Journal of the Elisha Mitchell Scientific Society* 54: 176–77.
Boyce, S. G. 1953. The salt spray community. Ph.D. dissertation, North Carolina State College.
———. 1954. The salt spray community. *Ecological Monographs* 24: 29–67.
Brown, D. M. 1941. Vegetation of Roan Mountain: a phytosociological and successional study. *Ecological Monographs* 11: 61–97.
Brown, R. 1978. Teacher honored after 40 years. *News and Observer*, 28 May.
Buhr, H. 1964. *Bestimmungstabellen der Gallen (Zoo- und Phytocecidien) an Pflanzen Mittel- und Nordeuropas*. 2 volumes. Jena: Gustav Fischer.
Cappell, E. D. 1953. The genus *Scirpus* in North Carolina. M.S. thesis, North Carolina State College.
Carpenter, W. L. 1978. *Let The People Know: A History of Agricultural Information Activities at North Carolina State University, 1879–1978*. Raleigh: North Carolina Agricultural Extension Service and North Carolina Agricultural Experiment Station.

Carpenter, W. L., and D. W. Colvard. 1987. *Knowledge Is Power: A History of the School of Agriculture and Life Sciences at North Carolina State University, 1877–1984*. Raleigh: North Carolina State University.

Clements, F. E. 1916. *Plant Succession: An Analysis of the Development of Vegetation*. Washington, D.C.: Carnegie Institution of Washington Publication 242.

———. 1936. Nature and structure of the climax. *Journal of Ecology* 24: 252–84.

The College Blue Book. 1977. 16th edition, Vol. 1. New York: Macmillan Information.

Committee on the Teaching of Botany in American Colleges and Universities. 1938. *An Exploratory Study of the Teaching of Botany in the Colleges and Universities of the United States*. Botanical Society of America.

Connor, L., Jr. 1938. Students pay tribute to Mrs. Edna M. Wells. *News and Observer*, 8 February.

Cook, M. T. 1917a. Current literature. Notes for students. Cecidology. *Botanical Gazette* 63: 158.

———. 1917b. Current literature. Notes for students. Galls. *Botanical Gazette* 64: 345–46.

———. 1923. The origin and structure of plant galls. *Science* 57: 6–14.

Cooper, A. W. 1979a. Bertram Whittier Wells, 1884–1978. *ASB Bulletin* 26: 123–24.

———. 1979b. Bertram Whittier Wells. *Veröffentlichungen des Geobotanischen Institutes der ETH, Stiftung Rübel, Zürich* 68: 7–10.

Core, E. L. 1968. The Natural Gardens of North Carolina. *Castanea* 33: 83–84.

Couch, W. T. 1946. *Books from Chapel Hill*. Chapel Hill: University of North Carolina Press.

Coville, F. C. 1937. Improving the wild blueberry. *United States Department of Agriculture Yearbook*: 559–74.

Cowdrey, A. E. 1983. *This Land, This South: An Environmental History*. Lexington: University Press of Kentucky.

Crafton, W. M. 1932. The old field prisere: an ecological study. M.S. thesis, North Carolina State College.

Craven, W. F., and J. L. Cate, eds. 1955. *Men and Planes: The Army Air Forces in World War II*. Vol. 6. Chicago: University of Chicago Press.

Cromelin, L. M. 1933. The Natural Gardens of North Carolina, by B. W. Wells. *American Forests* 39: 327.

Crow, A. B. 1973. Use of fire in southern forests. *Journal of Forestry* 71: 629–32.

Cunningham, B., D. B. Anderson, and Z. P. Metcalf. 1938. Mrs. Edna Metz Wells. *Journal of the Elisha Mitchell Scientific Society* 54: 175.

Dachnowski, A. P. 1912. Peat deposits of Ohio: their origin, formation, and uses. *Geological Survey of Ohio Bulletin* 16.

Daniels, J. 1932. Splendid work on state flora. *News and Observer*, 25 December.

Darrow, G. M. 1960. Blueberry breeding. Past, present, future. *American Horticultural Magazine* 39(1): 14–33.

Davis, D. 1940. Ecological escapade. *Pinetum*, Journal of Forestry of North Carolina State College, pp. 46–47, 59.
Davis, W. M. 1975. N.C. botanist lives to see movie of memorable book. *Greensboro Daily News*, 8 December.
———. 1976. Our "natural gardens" are now on film. *State* 43(9): 23–24, 36.
Deans, E. V., Jr. 1936. The Venus flytrap. *Nature* 28 (July): 13.
deLaubenfels, M. W. 1949. *Life Science*. 4th edition. New York: Prentice-Hall.
Denmark, L. P. 1934. The McLean murals. *News and Observer*, 14 December.
Dickey, M. G. 1910. Meetings of the Biological Club. April 11, 1910. *Ohio Naturalist* 10: 192.
———. 1911. Meetings of the Biological Club. October 3, 1910. *Ohio Naturalist* 11: 272.
di Stefano, M. 1967. Lineamente cecidologici di un maestro (a 4 mesi dalla morte del Prof. Alessandro Trotter). *Marcellia* 34: 119–33.
———. 1968. Elenco completo delle monografie e delgi studi cecidologici del Trotter. *Marcellia* 35: 3–44.
Duncan, W. H., and L. E. Foote. 1975. *Wildflowers of the Southeastern United States*. Athens: University of Georgia Press.
Dundon, T. R. 1962. Multinucleate giant cell formation in a *Pachypsylla* gall on *Celtis*. *American Journal of Botany* 49: 800–805.
Earley, L. S. 1989. The natural gardens of B. W. Wells. *Wildlife in North Carolina* 53(2): 18–23.
Ecological Society of America. 1937. Membership list. *Bulletin of the Ecological Society of America* 18(4): 66, 68.
———. 1942. *Directory*.
Egler, F. E. 1951. A commentary on American plant ecology based on the textbooks of 1947–1949. *Ecology* 32: 673–95.
———. 1977. *The Nature of Vegetation, Its Management and Mismanagement: An Introduction to Vegetation Science*. Norfolk and Bridgewater, Conn.: Aton Forest and Connecticut Conservation Association.
Elias, T. S. 1970. The genera of Ulmaceae in the southeastern United States. *Journal of the Arnold Arboretum* 51: 18–40.
Estes, B. 1986. Plants power architect of botanical garden. *News and Observer*, 23 June.
Everett, M. [H. Neil]. 1901. *Complete Life of William McKinley and Story of His Assassination*. N.p.: The author.
Federal Writers' Project. 1939. *A Guide to the Old North State*. Chapel Hill: University of North Carolina Press.
Felt, E. P. 1918. Key to American gall insects. *New York State Museum Bulletin* 200.
———. 1940. *Plant Galls and Gall Makers*. Ithaca, N.Y.: Comstock Publishing Company.
Flory, W. P. 1987. A 50-year historical perspective, 1937–1987, of the Association of Southeastern Biologists. *ASB Bulletin* 34(2)(supplement): 1–21.
Ford, G. S., F. L. McVey, and G. A. Works. 1932. Report to the North Carolina Commission on University Consolidation. In *Report of the Commission on University Consolidation*, pp. 9–100. Raleigh: Commission on University Consolidation.

Fritsch, F. E. 1945. *The Structure and Reproduction of the Algae*. Vol. 2. Cambridge: Cambridge University Press.

Fuller, G. D. 1933a. Nature's gardens and man's gardens. *Botanical Gazette* 94: 830–32.

———. 1933b. Plant ecology for the layman. *Ecology* 14: 407–8.

Furniss, N. F. 1963. *The Fundamentalist Controversy, 1918–1931*. Hamden, Conn.: Archon Books.

Gagné, R. J. 1983. *Celticecis* (Diptera: Cecidomyiidae), a new genus for gall makers on hackberries, *Celtis* spp. (Ulmaceae). *Proceedings of the Entomological Society of Washington* 85: 435–38.

———. 1984. The geography of gall insects. In *Biology of Gall Insects*, edited by T. N. Ananthakrishnan, pp. 305–22. London: Edward Arnold.

Galletta, G. J. 1975. Blueberries and cranberries. In *Advances in Fruit Breeding*, edited by J. Janick and J. N. Moore, pp. 154–96. West Lafayette, Ind.: Purdue University Press.

Garden Club of North Carolina. 1931–44. *Yearbook*. Raleigh: Garden Club of North Carolina, Inc.

Gatewood, W. B., Jr. 1960a. North Carolina's role in the establishment of the Great Smoky Mountains National Park. *North Carolina Historical Review* 37: 165–84.

———. 1960b. *Eugene Clyde Brooks: Educator and Public Servant*. Durham, N.C.: Duke University Press.

———. 1966. *Preachers, Pedagogues, and Politicians: The Evolution Controversy in North Carolina, 1920–1927*. Chapel Hill: University of North Carolina Press.

———, ed. 1969. *Controversy in the Twenties: Fundamentalism, Modernism, and Evolution*. Nashville: Vanderbilt University Press.

Gersmehl, P. J. 1970. A geographical approach to a vegetation problem: the case of the southern Appalachian grassy balds. Ph.D. thesis, University of Georgia.

Gleason, H. A., and A. Cronquist. 1964. *The Natural Geography of Plants*. New York: Columbia University Press.

Godfrey, R. K. 1938. The Compositae of Wake County, North Carolina, in time and space. M.S. thesis, North Carolina State College.

———. 1953. William Basil Fox. *Journal of the Elisha Mitchell Scientific Society* 69: 77.

Golley, F. B., ed. 1977. *Ecological Succession*. Stroudsburg, Pa.: Dowden, Hutchinson, and Ross.

Green, A. 1986. Botanist seeks to preserve area near lake. *News and Observer*, 2 January.

Green, C. H. 1974. On the Indian trail tree. *News and Observer*, 3 March.

Greene, W. F., and H. L. Blomquist. 1953. *Flowers of the South: Native and Exotic*. Chapel Hill: University of North Carolina Press.

Griggs, R. F. 1941. [Tribute to J. H. Schaffner.] In Professor John Henry Schaffner, by A. E. Waller. *Ohio Journal of Science* 41: 253–86.

———. 1961. *We Two Together*. Pittsburgh: Boxwood Press.

Gurney, A. B. 1961. The entomological works of Bentley B. Fulton. *Proceedings of the Entomological Society of Washington* 66: 151–59.

Haldane, J. B. S. 1923. If you were alive in 2123 A.D. *Century Magazine* 106: 549–66.
Hale, H. 1948. *University of Arkansas*. Fayetteville: University of Arkansas Alumni Association.
Halstead, M. 1901. *The Illustrious Life of William McKinley Our Martyred President*. N.p.: The author.
Hardin, J. W. 1984. Botany at Mountain Lake Biological Station: a brief history. *ASB Bulletin* 31(1): 3–6.
Harper, R. M. 1913. *Economic Botany of Alabama. Part 1, Monograph 8, Geological Report on Forests* 25: 1.
Harrelson, J. W. 1944. North Carolina State College. Report of the Dean of Administration. In *The University of North Carolina, A Ten Year Review*, n.p. Chapel Hill: University of North Carolina Press.
Harshberger, J. W. 1911. *Phytogeographic Survey of North America*. Leipzig: Wilhelm Engelmann.
Heltzel, J. B. 1936. An ecological fantasy. *Pinetum*, Journal of Forestry of North Carolina State College, pp. 39–42.
Henderson, I. B. 1935. Breaking an agreement. *News and Observer*, 25 January.
Hodkinson, I. D. 1984. The biology and ecology of the gall-forming Psylloidea (Homoptera). In *Biology of Gall Insects*, edited by T. N. Ananthakrishnan, pp. 59–77. London: Edward Arnold.
Iden, S. F. 1912. *Edenton Street Methodist Sunday School Raleigh, N.C.; Historical Sketch Commemorating the Opening of the New Sunday School Building, April 28, 1912*. Raleigh: Edwards and Broughton Printing Company.
———. 1920. The Raleigh pageant. *Training School Quarterly* 8(1): 15–19.
———. 1926. North Carolina makes another claim to greatness. Nature study trip reveals wonders of coastal plain region. *Raleigh Times*, 24 July.
Johnson, T. W., Jr. 1969. Henry John Oosting. *Journal of the Elisha Mitchell Scientific Society* 85: 108.
Jones, T. 1989. Raleigh artist J. A. McLean dies at age 84. *News and Observer*, 1 March.
Journal. 1977–82. Raleigh: North Carolina State University.
Justice, W. S., and C. R. Bell. 1968. *Wild Flowers of North Carolina*. Chapel Hill: University of North Carolina Press.
Kandasamy, C. 1982. An uncommon psyllid gall of *Ficus bengalensis*. *Current Science* 51: 713–14.
Kinsey, A. C. 1920. Phylogeny of cynipid genera and biological characteristics. *Bulletin of the American Museum of Natural History* 42: 357a–357c, 358–402.
———. 1930. The gall wasp genus *Cynips*. A study in the origin of species. *Indiana University Studies* 16, Numbers 84, 85, 86.
Kirk, J. S., W. A. Cutter, and T. W. Morse, eds. 1936. *Emergency Relief in North Carolina: A Record of the Development and Activities of the North Carolina Emergency Relief Administration, 1932–1935*. Raleigh: North Carolina Emergency Relief Commission.
Kirkman, W. B., and J. R. Ballington. 1985. 'Wells Delight' and 'Bloodstone' creeping blueberries. *HortScience* 20: 1138–40.
Knight, E. W. 1927. Monkey or mud in North Carolina? *Independent* 118: 515–16, 523.

LaFollette, C. T. 1935. Home economics in New College. *Journal of Home Economics* 27: 342–47.

Lawrence, D. B. 1979. Resolution of respect, William Skinner Cooper (1884–1978). *Bulletin of the Ecological Society of America* 60(1): 18–19.

Lawrence, E. 1966. Pilot Mountain visited by wild flower society. *Charlotte Observer*, 16 October.

———. 1968. There are 11 major plant areas in N.C. *Charlotte Observer*, 10 March.

Lewis, I. F., and L. Walton. 1964. Gall formation on leaves of *Celtis occidentalis* L. resulting from material injected by *Pachypsylla* sp. *Transactions of the American Microscopical Society* 83: 62–78.

Lewis, N. B. 1926. North Carolina. *American Mercury* 8(29): 36–42.

Lieth, H., E. Landolt, and R. K. Peet, eds. 1979, 1980, 1981. Contributions to the knowledge of flora and vegetation in the Carolinas. Proceedings of the 16th International Phytogeographical Excursion (IPE), 1978, through the southeastern United States. *Veröffentlichungen des Geobotanischen Institutes der ETH, Stiftung Rübel, Zürich* 68, 69, 77.

Linder, S. C. 1963. William Louis Poteat and the evolution controversy. *North Carolina Historical Review* 40: 135–38, 155–56.

———. 1966. *William Louis Poteat: Prophet of Progress*. Chapel Hill: University of North Carolina Press.

Liser y Trelles, C. A., and C. C. Molle. 1945. Estructura anatomica de filocecidias neotropicas. *Lilloa* 11: 153–87.

Litt, S. 1985. Artist aided in birth of RLT. *News and Observer*, 31 July.

Lockmiller, D. A. 1939. *History of the North Carolina State College of Agriculture and Engineering of the University of North Carolina*. Raleigh: General Alumni Association of the North Carolina State College.

———. 1942. *The Consolidation of the University of North Carolina*. Raleigh, Chapel Hill, and Greensboro: University of North Carolina.

Lynch, B. 1952. .22-cal. bullet kills NCS professor. William Basil Fox's death probe pushed. *Raleigh Times*, 13 November.

McCalla, D. R., M. K. Genthe, and W. Hovanitz. 1962. Chemical nature of an insect gall growth factor. *Plant Physiology* 37: 98–103.

McCray, A. H. 1909. Meetings of the Biological Club, November 2, 1908. *Ohio Naturalist* 9: 455–56.

McIntosh, R. P. 1976. Ecology since 1900. In *Issues and Ideas in America*, edited by B. J. Taylor and T. J. White, pp. 353–72. Norman: University of Oklahoma Press.

———. 1985. *The Background of Ecology: Concept and Theory*. Cambridge: Cambridge University Press.

McMenamin, J. P. 1940. The leaf anatomy of southeastern shrub-bog plants. M.S. thesis, North Carolina State College.

McPherson, H. 1976. *High Pointers of High Point*. High Point, N.C.: Chamber of Commerce of High Point.

Maresquelle, H. J., and J. Meyer. 1965. Physiologie et morphogenèse galles d'origine animale (zoocécidies). In *Handbuch der Pflanzenphysiologie*, edited by W. Ruhland, Vol. 15, part 2, pp. 280–329. Berlin: Springer-Verlag.

Mark, A. F. 1958. The ecology of the southern Appalachian grass balds. *Ecological Monographs* 28: 293–336.
Marshall, J. 1941. Opinions political and otherwise. *Raleigh Times*, 25 September.
Martin, A. C. 1924. An ontogenetic study of the gall *Phylloxera caryaeseptem* Shimer. M.S. thesis, North Carolina State College.
Martin, A. C., and W. D. Barkley. 1961. *Seed Identification Manual*. Berkeley: University of California Press.
Meyer, B. S. 1983. Botany at the Ohio State University: the first 100 years. *Bulletin of the Ohio Biological Survey* 6(2): 1–177.
Mountain Lake Biological Station. 1939, 1950. *Catalog*. Charlottesville: University of Virginia.
Morrison, J. L. 1971. *Governor O. Max Gardner: A Power in North Carolina and New Deal Washington*. Chapel Hill: University of North Carolina Press.
National Union Catalog. 1979. *Pre-1956 Imprints* 655: 236.
News and Observer. 1922–92. Raleigh, N.C.
North Carolina Academy of Science. 1924. Proceedings of the twenty-third annual meeting, May 2–3. *Journal of the Elisha Mitchell Scientific Society* 40: 93–112.
———. 1926. Proceedings of the twenty-fifth annual meeting, April 30–May 1. *Journal of the Elisha Mitchell Scientific Society* 42: 1–18.
North Carolina Agricultural Experiment Station. 1938. *Annual Report*: 49.
North Carolina Agricultural Research Service. 1984. Blueberry lines for ground cover. *Research Perspectives* 3(3): 15.
North Carolina Agriculture and Industry. 1923–26. Raleigh: North Carolina State College.
North Carolina Department of Agriculture. 1926–28 through 1938–40. *Biennial Report*. Raleigh: North Carolina Department of Agriculture.
North Carolina Library Commission. 1921. *Sixth Biennial Report, 1919–1920*. Raleigh: North Carolina Library Commission.
North Carolina Nature Conservancy. 1983, 1985. *Newsletter*.
North Carolina State College Catalog. 1919–54. Raleigh: North Carolina State College of Agriculture and Engineering.
North Carolina State University Student Center. 1987. *Seeing an Idea: Gifts and Acquisitions since 1982*. Raleigh: North Carolina State University.
North Carolina Wild Flower Preservation Society. 1955–72. *Newsletter*.
North Carolina Wildlife Federation. 1979. Dr. Bertram Whittier Wells. Governor's Award Conservation Education. *Friend O'Wildlife* 26(3): 5.
Norton, T. A. 1969. Growth form and environment in *Saccorhiza polyschides*. *Journal of the Marine Biological Association of the United Kingdom* 49: 1025–45.
Nygard, M. 1989. Seven million years of Ziegle's Rock and the legacy of B. W. Wells, who bought it and loved it. *Leader, Magazine of the Triangle*, 30 November, pp. 12–13.
Ohio State Academy of Science. 1911–30. Annual report. *Proceedings of the Ohio State Academy of Science*.
O'Keef, H. 1955. Tar heel of the week. B. W. Wells. *News and Observer*, 30 January.

Oosting, H. J. 1945. Tolerance to salt spray of plants of coastal dunes. *Ecology* 26: 85–89.
———. 1948. *The Study of Plant Communities: An Introduction to Plant Ecology*. San Francisco: W. H. Freeman and Company.
———. 1954. Ecological processes and vegetation of the maritime strand in the southeastern United States. *Botanical Review* 20: 226–62.
———. 1956. *The Study of Plant Communities: An Introduction to Plant Ecology*. 2d edition. San Francisco: W. H. Freeman and Company.
Oosting, H. J., and W. D. Billings. 1941. The distribution of dune vegetation and the effect of salt spray. *Journal of the Elisha Mitchell Scientific Society* 57: 187.
———. 1942. Factors affecting vegetational zonation on coastal dunes. *Ecology* 23: 131–42.
Ousley, C. 1935. History of the Agricultural and Mechanical College of Texas. *Bulletin of the Agricultural and Mechanical College of Texas*, 4th series, 6(8): 1–172.
Overton, S. 1987. The natural. For 97 years, Charlotte Hilton Green has followed life down unbeaten paths. *News and Observer*, 23 April.
Painter, A. 1977. Speaking of ecology: he wrote the book. *State* 55(3): 24.
Papenfuss, G. F. 1951. Phaeophyta. In *Manual of Phycology*, edited by G. M. Smith, pp. 119–58. New York: Ronald Press.
Parkins, E. J. 1982. Pioneer ecologist's farm will become nature study area. *Raleigh Times*, 25 May.
Payne, P. 1977. Riverbend botanist: Dr. Bertram Wells. *Stater, North Carolina State University's Magazine for Alumni* 49(6): 14–15.
Pollard, D. G. 1973. Plant penetration by feeding aphids (Hemiptera, Aphidoidea): a review. *Bulletin of Entomological Research* 62: 631–714.
Primack, R. B., and R. Wyatt. 1975. Variation and taxonomy of *Pyxidanthera* (Diapensiaceae). *Brittonia* 27: 115–18.
Prouty, W. F. 1921. A more phenomenal shoot. *Science* 54: 170.
Quillen, M. 1992. Charlotte H. Green, naturalist, author, 102. *News and Observer*, 27 March.
Radford, A. E., J. E. Adams, C. R. Bell, V. A. Greulach, and L. S. Olive. 1973. Memorial to Henry Roland Totten. *Journal of the Elisha Mitchell Scientific Society* 89: 262–63.
Radford, A. E., H. E. Ahles, and C. R. Bell. 1968. *Manual of the Vascular Flora of the Carolinas*. Chapel Hill: University of North Carolina Press.
Raleigh Times. 1924–82. Raleigh, N.C.
Raper, C. F. 1954. With Wells along the Waccamaw. *Pinetum*, Journal of Forestry of North Carolina State College, p. 47.
Reagan, A. E. 1987. *North Carolina State University, A Narrative History*. Raleigh: North Carolina State University Foundation and North Carolina State University Alumni Association.
Reid, J. B. 1952. Hitting the pamlico trail or Dr. Wells, I presume! *Pinetum*, Journal of Forestry of North Carolina State College, pp. 43–46.
Renfro, J. 1951. Ecology for moderns. *Pinetum*, Journal of Forestry of North Carolina State College, pp. 43–44.
Reynolds, J. D. 1968. Morphological studies in the Diapensiaceae. I. Chro-

mosome number and microspore development in *Pyxidanthera brevifolia* Wells. *Bulletin of the Torrey Botanical Club* 95: 653–56.
Robertson, S. 1924. Reorganization N.C. State would unify college work; best thought must direct. *News and Observer*, 20 January.
Robinson, B. P., ed. 1955. *The North Carolina Guide*. Chapel Hill: University of North Carolina Press.
Robson, J. W. 1937. *A Guide to Columbia University, with Some Account of Its History and Traditions*. New York: Columbia University Press.
Roller, J. H. 1931. Ecological aspects of the sandhills. M.S. thesis, North Carolina State College.
Schneider, J. 1983. The camera collector. *Modern Photography*, June, pp. 64–65, 99–100.
Sears, P. B. 1914. The insect galls of Cedar Point and vicinity. *Ohio Naturalist* 15: 377–92.
Sharpe, B. 1965. The natural gardens. *State* 35(19): 16–17, 25.
Shelford, V. E., ed. 1926. *Naturalist's Guide to the Americas*. Baltimore: Williams and Wilkins Company.
Shelley, P. N. 1974. Dr. B. W. Wells made valuable contributions as teacher, scientist, author, and artist. *Journal* 5(5): 6.
Shimwell, D. W. 1971. *The Description and Classification of Vegetation*. Seattle: University of Washington Press.
Shipley, M. 1927. *The War on Modern Science: A Short History of the Fundamentalist Attacks on Evolution and Modernism*. New York: Alfred A. Knopf.
Simpson, D. 1982. On the lake. Plans for Falls Lake facilities offer wide choice of activities. *Raleigh Times*, 7 August.
Small, J. K. 1913. *Flora of the Southeastern United States*. 2d edition. New York: The author.
———. 1933. *Manual of the Southeastern Flora*. New York: The author.
Smathers, G. A. 1980. The anthropic factor in southern Appalachian bald formation. In *Status and Management of Southern Appalachian Mountain Balds*, edited by P. R. Saunders, pp. 18–38. Crossnore, N.C.: Southern Appalachian Research/Resource Management Cooperative.
Smith, R. C., and R. S. Taylor. 1953. The biology and control of the hackberry psyllids in Kansas. *Journal of the Kansas Entomological Society* 26: 103–15.
Snyder, J. R. 1980. Analysis of coastal plain vegetation, Croatan National Forest, North Carolina. *Veröffentlichungen des Geobotanischen Institutes der ETH, Stiftung Rübel, Zürich* 69: 40–113.
State College Record. 1924–36. Raleigh: North Carolina State College of Agriculture and Engineering.
Stick, D. 1985. *Bald Head: A History of Smith Island and Cape Fear*. Wendell, N.C.: Broadfoot Publishing Company.
Teale, E. W. 1951. *North with the Spring*. New York: Dodd, Mead, and Company.
Technician. 1924–82. Raleigh: Student newspaper of North Carolina State College.
Thompson, H. 1932. Report on the dismissal of Dean Carl C. Taylor. *Bulletin of the American Association of University Professors* 18(3): 224–32.

Tobey, R. C. 1981. *Saving the Prairies: The Life Cycle of the Founding School of American Plant Ecology*. Berkeley: University of California Press.
Totten, Mrs. H. R., L. Ballard, and Mrs. G. W. Little, eds. 1959. *The First Thirty-Four Years, 1925–1959: A Fact Finder of the Garden Club of North Carolina, Inc.* Raleigh: Garden Club of North Carolina, Inc.
Trotter, A. 1915. Bibliografia e recensioni. *Marcellia* 14: XIX.
———. 1916. Bibliografia e recensioni. *Marcellia* 15: IV.
———. 1920. Bibliografia e recensioni. *Marcellia* 17: XXII–XXIII.
———. 1922/1923. Intorno all' evoluzione morfologica delle galle. *Marcellia* 19: 120–47; 20: 67–86.
———. 1935. Bibliografia e recensioni. *Marcellia* 29: IV.
Troyer, J. R. 1986. Bertram Whittier Wells (1884–1978): a study in the history of North American plant ecology. *American Journal of Botany* 73: 1058–78.
———. 1987. B. W. Wells, Z. P. Metcalf, and the North Carolina Academy of Science in the evolution controversy, 1922–1927. *Journal of the Elisha Mitchell Scientific Society* 102: 43–53.
University of North Carolina at Greensboro. 1976. A garden graphic. *Alumni News* 64(2): 1.
Vanderlaan, E. C. 1925. *Fundamentalism versus Modernism*. New York: H. W. Wilson Company.
Vess, T. 1982a. Event honors artist; mural displayed in Student Center. *Technician*, 11 October.
———. 1982b. Artist presents nostalgic mural to State. *Technician*, 13 October.
W., S. B. 1967. Book larning. *State* 35(13): 16.
Walton, B. C. J. 1960. The life cycle of the hackberry gall-former, *Pachypsylla celtidis-gemma* (Homoptera: Psyllidae). *Annals of the Entomological Society of America* 53: 265–77.
Watson, G. 1964. Utopia and rebellion: the New College experiment. In *Innovation in Education*, edited by M. B. Miles, pp. 97–115. New York: Bureau of Publications, Teachers College, Columbia University.
West, F. 1948. Down to the sea. *Pinetum*, Journal of Forestry of North Carolina State College, pp. 49–50.
White, B. 1960. Wanted: an unspoiled savannah. *News and Observer*, 4 September.
Whitehead, D. R. 1972. Developmental and environmental history of the Dismal Swamp. *Ecological Monographs* 42: 301–15.
Whitford, L. A. 1929. The algae of Lake Raleigh: an ecological study. M.S. thesis, North Carolina State University.
Whittaker, R. H. 1953. A consideration of climax theory: the climax as a population and pattern. *Ecological Monographs* 23: 41–78.
———. 1962. Classification of natural communities. *Botanical Review* 28: 1–239.
Williams, C. F. 1924. The morphological ecology of savannah plants. I. *Campulosus aromaticus* (Walt.) Scribn. M.S. thesis, North Carolina State College.
Wilson, L. R. 1957. *The University of North Carolina, 1900–1930: The Making of a Modern University*. Chapel Hill: University of North Carolina Press.

———. 1960. *Harry Woodburn Chase*. Chapel Hill: University of North Carolina Press.

———. 1964. *The University of North Carolina under Consolidation, 1931–1963: History and Appraisal*. Chapel Hill: University of North Carolina Consolidated Office.

Zweigelt, F. 1931. *Blattlaugallen. Histogenetische und biologische Studien an Tetraneura- und Shizoneuragallen. Die Blattlaugallen im Dienste prinzipieller Gallenforschung*. Berlin: Verlagsbuchhandlung Paul Parey.

INDEX

Abbreviations

BWW	Bertram Whittier Wells
EMW	Edna Metz Wells
GCNC	Garden Club of North Carolina
MBW	Maude Barnes Wells
NCSC	North Carolina State College and University
NGNC	*The Natural Gardens of North Carolina*
WFPS	North Carolina Wild Flower Preservation Society